The language of p

The language of physics
A foundation for university study

John P. Cullerne
Head of Physics, Winchester College, Winchester

Anton Machacek
Head of Physics, Royal Grammar School, High Wycombe

OXFORD
UNIVERSITY PRESS

OXFORD
UNIVERSITY PRESS

Great Clarendon Street, Oxford OX2 6DP

Oxford University Press is a department of the University of Oxford.
It furthers the University's objective of excellence in research, scholarship,
and education by publishing worldwide in

Oxford New York

Auckland Cape Town Dar es Salaam Hong Kong Karachi
Kuala Lumpur Madrid Melbourne Mexico City Nairobi
New Delhi Shanghai Taipei Toronto

With offices in

Argentina Austria Brazil Chile Czech Republic France Greece
Guatemala Hungary Italy Japan Poland Portugal Singapore
South Korea Switzerland Thailand Turkey Ukraine Vietnam

Oxford is a registered trade mark of Oxford University Press
in the UK and in certain other countries

Published in the United States
by Oxford University Press Inc., New York

© J.P. Cullerne and A.C. Machacek, 2008

The moral rights of the author have been asserted
Database right Oxford University Press (maker)

First Published 2008

Reprinted 2012, 2013 (twice)

All rights reserved. No part of this publication may be reproduced,
stored in a retrieval system, or transmitted, in any form or by any means,
without the prior permission in writing of Oxford University Press,
or as expressly permitted by law, or under terms agreed with the appropriate
reprographics rights organization. Enquiries concerning reproduction
outside the scope of the above should be sent to the Rights Department,
Oxford University Press, at the address above

You must not circulate this book in any other binding or cover
and you must impose the same condition on any acquirer

British Library Cataloguing in Publication Data

Data available

Library of Congress Cataloging in Publication Data

Data available

Typeset by Newgen Imaging Systems (P) Ltd., Chennai, India
Printed in Great Britain
on acid-free paper by
CPI Group (UK) Ltd, Croydon, CR0 4YY

ISBN 978-0-19-953379-4 (Hbk)
ISBN 978-0-19-953380-0 (Pbk)

5 7 9 10 8 6 4

Preface

Over the past decade we have seen a definite shift in knowledge base of our students. Presently, at pre-university level, mathematics seems to be less integrated with the study of physics and some of the most important topics are covered only in Further Mathematics courses. Results and techniques are often learnt in school as mathematical processes without much regard for the underlying principles. Hence, students often find it hard to build up a mathematical description of a physical system from scratch. This is an essential skill required for any undergraduate degree in physics or engineering.

You will notice that we have tried, where possible, to integrate the mathematics into the physics so that the reader is given a chance to see the physics unfold in the most appropriate language (mathematics). The reader is given ample opportunity to try out the language for themselves in workshop sections, which have been designed to show intermediate steps and results and to help the reader through some of the most conceptually difficult demonstrations. Fully worked solutions to all workshops are presented as an appendix to this book. There are also questions following each section that deal with principles studied in that section.

While we have not seen 'workshops' as such in other books, the idea is straightforward. When carpenters build things out of wood, they have to fashion them into pieces that have the right shape first. They do this in a workshop where they are surrounded by the right tools. Our workshops have been designed with this in mind – the students enter a section that is specifically designed to help them get to grips with a particular mathematical concept that will transform their understanding of the physics. As with the carpenter, the mathematics is 'fit for purpose' – we choose to convey the mathematical techniques with a strong reference to the context of physics.

The assumed knowledge base here is really only that of a standard pre-university course in mathematics (such as a single mathematics A-level in the UK). The physics is developed using the mathematics as a tool, and while pre-university study of physics is not assumed, we cover the necessary concepts quite rigorously, and previous study would be beneficial. The 'syllabus' is intended to form a convenient stepping stone between school and undergraduate study in the physical sciences or engineering. By requiring a good measure of problem solving (which itself requires a deeper understanding of concepts), it is possible to design questions that do not venture into university mathematics, but would nevertheless give most undergraduates a good run for their money.

Therefore we hope the present text will be used by first-year undergraduate students as they grapple (perhaps for the first time) with their physics or engineering written in the language of mathematics. We also would hope that it would be used by not a few dedicated pre-university students in the run up to their first-year undergraduate course. And it may even be of interest to any physical scientists who need or are compelled to espy a little of the fundamental mathematics that lies behind their physics.

JPC and ACM
Winchester and High Wycombe, 2008

To the Student ...

One of us once heard two senior school students muttering in the back of the class, 'Why is there so much calculus in our maths course? It's not as if it's any use ...' Once those students went to university to study engineering, they discovered that a knowledge of calculus is as vital as knowing that 9 is bigger than 5, and that there was precious little useful information written *without* calculus. After all, the true language of physics is not English. It is mathematics. The aim of this book is to help you develop fluency in the true language of your undergraduate subject by explaining the physics you know in terms of mathematics, and showing you how this enables you to solve a wider range of problems at a more advanced level.

To get the most out of this book you need to have studied, or be studying, a pre-university course in mathematics (such as single mathematics A-level in the UK). It will help if you have also studied physics at this level and/or part of an extra mathematics course, but these are not assumed. If you have studied physics, we hope that this book helps you 'bridge the gap' between two disciplines taught so separately at school. If you have not studied physics, we hope that this book gives you a mathematically oriented introduction to the subject.

Practice is vital to developing fluency in any language. Accordingly, there are many problems to be worked through. Harder problems are marked + or ++. The most essential exercises are in the form of 'workshops' which lead you through a new technique or concept by the hand. Full solutions to these workshops are included as an appendix to this book, and we hope that you will use them regularly (looking in the back is not cheating, it is learning). If you want a summary of what you have learned, please look through the relevant part of Chapter 8 where you will find a summary of the equations used – which is probably the best way of summarizing the content of each chapter in the book. With any complex calculation, we advise you to work in terms of the parameters represented by letters (t, s, E, γ, etc.), and only to substitute numbers once you are sure you have the correct algebraic expression.

In addition to undergraduates, we hope that some of our readers will be students still at school wishing to enrich their understanding and gain a better taste for how subjects such as physics are presented at university. If you are working by yourself, you will probably find this quite tough, but either of us would be delighted to hear from you if you require further help. We would encourage any students working without a teacher to make use of the Solutions Manual available on the publisher's website. We are, of course, deeply grateful to any student, lecturer or teacher who writes to us (through our publisher) with feedback.

Einstein referred to nature as subtle, not malicious. Some of the techniques may seem malicious, that is unnecessarily complicated, as indeed the theories of nature often appear. Our aim, though, is that you should come through the suspicion of malice to an appreciation of the subtlety of physical thought, and that one day this will help you appreciate the mathematical beauty of nature herself.

JPC and ACM
Winchester and High Wycombe, 2008

Dedications and Acknowledgements

There are many people who have helped us write this book and encouraged us during our travels in physics and mathematics. Space precludes us mentioning them all, but we have particularly appreciated the discussions we have had (and the support we have received) from the Physics Departments of Winchester College and the Royal Grammar School – teachers, technicians, and students alike.

We are grateful for the encouragement given to us by Oxford University Press to write this book – their guidance has been warm, rigorous, and professional, and we could not have wished for more from a publisher. It is only right for us also to acknowledge publically that we have greatly enjoyed working together as authors, and that this project has not only deepened our friendship but also developed our enjoyment and appreciation of physics immeasurably.

We owe a still deeper debt of gratitude to our beloved wives Kay and Helen for their love and support in a whole plethora of ways – including their commitment to this book. Kay and Helen have shared our thoughts, helped us in our labour, been patient when we have been busy and encouraged us continually. We are fortunate indeed to have such companions on the journey of life.

As these words leave our hands to go to the publisher, and thence to you as our reader, John and Kay dedicate them to their children, and Anton and Helen dedicate them to the glory of God.

JPC and ACM
Winchester & High Wycombe, 2007

Contents

1	**Linear mechanics**		1
	1.1 Kinematics		1
	1.1.1 The law of falling bodies		1
	1.1.2 The kinematics of falling bodies		4
	1.1.3 Workshop: Simple differential equations		9
	1.1.4 The kinematics of a projectile		11
	1.1.5 Workshop: Motion on the surface of a smooth inclined plane		14
	1.1.6 Adding and subtracting vectors		15
	1.2 Dynamics		17
	1.2.1 Newton's laws		17
	1.2.2 The principle of relativity		21
	1.2.3 Impulse and impulsive forces		22
	1.2.4 Workshop: The conservation of linear momentum		23
	1.2.5 The law of falling bodies		25
	1.2.6 Workshop: Newton and the apple		26
	1.3 Conclusion		27
2	**Fields**		29
	2.1 Introduction and field strength		29
	2.2 Workshop: Motion in a uniform field in one dimension		30
	2.3 Workshop: Scalar product of vectors		32
	2.4 Workshop: Motion in a uniform field in three dimensions		34
	2.5 Non-uniform fields		36
	2.6 Workshop: Evaluating line integrals		37
	2.7 Potential gradients		39
	2.8 Setting up a field		44
	2.8.1 Workshop: The electrostatic field surrounding a charged wire		47
	2.8.2 Electrostatic charge in a parallel plate capacitor		48
	2.8.3 Gravitational fields inside planets		50
	2.8.4 Formalizing the notation		51
	2.9 Conclusion		53

xii Contents

3 Rotation — 54
- 3.1 Rotational kinematics and dynamics — 54
  - 3.1.1 Kinematics on a circular path — 54
  - 3.1.2 Workshop: Rotated coordinate systems and matrices — 56
  - 3.1.3 Workshop: Rotating vectors and the vector product — 58
  - 3.1.4 Angular velocity — 59
  - 3.1.5 Workshop: Vector triple product — 61
  - 3.1.6 Acceleration vectors in rotating frames — 62
  - 3.1.7 'Fictitious force': Centrifugal and Coriolis forces — 64
- 3.2 Orbits — 66
  - 3.2.1 The Kepler problem — 66
  - 3.2.2 Kepler's first law and properties of $(d^2/dt^2)\mathbf{r}$ — 70
  - 3.2.3 Workshop: Kepler's second law — 74
  - 3.2.4 Workshop: Kepler's third law — 75
- 3.3 Conclusion — 77

4 Oscillations and waves — 78
- 4.1 Describing an oscillation — 78
  - 4.1.1 Workshop: Simple harmonic motion — 80
- 4.2 Workshop: Introducing complex numbers — 81
- 4.3 Describing an oscillation using complex numbers — 84
- 4.4 Workshop: Damped oscillators — 85
- 4.5 Describing a wave in one dimension — 86
- 4.6 Interference – a brief introduction — 87
- 4.7 Workshop: The wave equation — 89
- 4.8 A wave on a string — 89
- 4.9 Energy content of a wave — 91
- 4.10 Impedance matching — 92
- 4.11 Describing waves in three dimensions — 94
  - 4.11.1 Plane waves — 94
  - 4.11.2 Spherical waves — 95
  - 4.11.3 Workshop: Stellar magnitudes — 96
- 4.12 Conclusion — 97

5 Circuits — 98
- 5.1 Fundamentals — 98
  - 5.1.1 Electric current — 99
  - 5.1.2 Electric potential — 100
  - 5.1.3 Workshop: Using voltage to solve simple circuit problems — 101
  - 5.1.4 Ohm's law and resistance — 101
- 5.2 Direct current circuit analysis — 102
  - 5.2.1 Analysis using fundamental principles — 103
  - 5.2.2 Method of loop currents — 104

5.3		Introducing alternating current	105
	5.3.1	Resistors	106
	5.3.2	Power in a.c. circuits and rms values	106
	5.3.3	Capacitors	108
	5.3.4	Inductors	109
	5.3.5	Sign conventions	109
	5.3.6	Phasor methods in a.c. analysis	109
5.4		Alternating current circuit analysis	111
	5.4.1	Analysis using impedances	112
	5.4.2	Analysis using a phasor	114
5.5		Conclusion	115

6 Thermal physics 116

6.1		The conservation of energy: The first law	116
6.2		The second law	117
6.3		Carnot's theorem	118
	6.3.1	Heat engines and fridges	118
	6.3.2	Thermodynamic temperature	121
	6.3.3	Efficiency of a heat engine	122
6.4		Entropy	123
	6.4.1	Reversible processes	123
	6.4.2	Irreversible processes and the second law	124
	6.4.3	Restatement of first law	125
6.5		The Boltzmann law	125
	6.5.1	Workshop: Atmospheric pressure	125
	6.5.2	Velocity distribution of molecules in a gas	126
	6.5.3	Workshop: Justification of Boltzmann law	127
6.6		Perfect gases	129
	6.6.1	Heat capacity of a perfect gas	130
	6.6.2	Pumping heat	131
6.7		Conclusion	135

7 Miscellany 136

7.1	Workshop: Setting up integrals	136
7.2	Workshop: Logarithms	138
7.3	Workshop: Rockets and stages	140
7.4	Workshop: Unit conversion	143
7.5	Workshop: Dimensional analysis	144
7.6	Workshop: Error analysis	147
7.7	Workshop: Centres of mass	150
7.8	Workshop: Rigid body dynamics	152
7.9	Workshop: Parallel axes theorem	155

7.10	Workshop: Perpendicular axes theorem	157
7.11	Workshop: Orbital energy and orbit classification	159

8 Summary of equations 162
8.1	Linear mechanics	162
8.2	Fields	163
8.3	Rotation	165
8.4	Waves	167
8.5	Circuits	169
8.6	Thermal physics	170

Workshop solutions 172
Chapter 1	172
Chapter 2	178
Chapter 3	181
Chapter 4	188
Chapter 5	197
Chapter 6	198
Chapter 7	203

Index 223

1
Linear mechanics

Kinematics is the study of motions within a framework of three-dimensional (3-D) space which are realized in the course of time. This study is made independently of the physical laws of these motions. *Dynamics* is the study of the physical laws of motion. It seems absolutely logical to study the different kinds of motion in space, before considering the reasons and according to what laws such and such a motion occurs in such and such a circumstance. In this chapter we will be following this quite traditional line of storytelling, but we will do so as a development of ideas rather than a compartmentalizing of the subject matter.

1.1 Kinematics

1.1.1 The law of falling bodies

From experiments it is possible to infer a simple general rule: *the motion of free fall is universally the same, independent of the size and material of the body.* The effect of the air on falling objects masks this general rule sometimes, so this seems a remarkable fact which people can find surprising. Further study of free fall reveals more than just the qualitative rule – a beautiful and simple pattern seems to emerge from the fall of an object under the influence of gravity alone (Figure 1.1).

If a body is released from rest, it falls a distance c in the first unit of time, then in the next unit of time the body will fall $3c$, then in the next unit it will fall $5c$, and so on. In successive units of time, the body falls distances that are *odd number* multiples of the distance fallen in the first unit. The total distance fallen from the point of release is then going to go as multiples of c following the perfect squares: $1, 1+3=4, 1+3+5=9$, and so on.

Therefore the total distance fallen (say s) can be conveniently represented as

$$s = ct^2, \qquad (1.1)$$

where t is the total time elapsed from the point of release. This simple relationship has been dubbed *the law of falling bodies*. Expression (1.1) holds no matter what interval of time Δt^* is chosen. This of course means that (1.1) describes a smooth curve when

---

*The Greek letter Δ or δ seems to crop up a lot in physics texts. This is not just to frighten people away with mathematical symbols. There is really a good reason for it. When it does appear, Δ or δ will always be followed by another letter, e.g. Δt or δt. Sometimes, in physics we wish to write a symbol for a change or step in a quantity rather than a particular value of a quantity. For example, we might want to use y to describe the y-coordinate of a point, but we might want to use Δy to

2 Linear mechanics

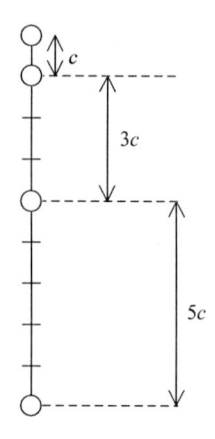

Fig. 1.1

plotted on Cartesian axes with s on the ordinate (vertical axis) and t on the abscissa (horizontal axis).

Let us now see what we can get out of (1.1) using the elementary concept:

$$\text{Average speed} = \text{distance} \div \text{time}. \tag{1.2}$$

Figure 1.2 depicts a small portion of the curve (1.1). The point $s(t)$ is the value of (1.1) evaluated at time t, and the point $s(t + \Delta t)$ is (1.1) evaluated at a later time $t + \Delta t$. Using (1.2) we can easily see that the average speed, v, of the body in free fall between times t and $t + \Delta t$ is just

$$v = \frac{s(t + \Delta t) - s(t)}{\Delta t}. \tag{1.3}$$

Using (1.1) we can see that

$$s(t + \Delta t) - s(t) = c(t + \Delta t)^2 - ct^2 = 2ct \cdot \Delta t + c\Delta t^2, \tag{1.4}$$

so (1.3) becomes

$$v = 2ct + c\Delta t. \tag{1.5}$$

Now the interesting thing about (1.5) is that as we make Δt smaller and smaller the point $s(t + \Delta t)$ gets closer and closer to the point $s(t)$, and in the limit when $\Delta t = 0$ we see that $s(t + \Delta t) = s(t)$ and (1.5) becomes

$$v = 2ct. \tag{1.6}$$

The reason why this quantity v remains finite even though both $s(t + \Delta t) - s(t)$ and Δt tend to 0 is that (1.3) is actually the gradient of the chord cutting the curve

describe the change or step in y-coordinate when a particle moves between two points separated along the y-axis. The lower case delta δ is used when we want to describe a very small change, so that the related quantities remain almost constant over the change.

Distance

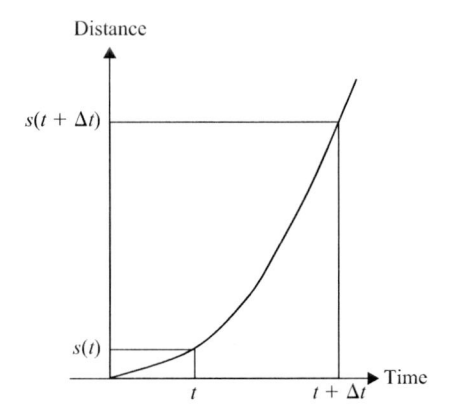

s(t + Δt)

s(t)

t   t + Δt   Time

Fig. 1.2

at points $s(t+\Delta t)$ and $s(t)$, and as $\Delta t \to 0$ this chord becomes a tangent to the curve at the point $s(t)$. The expression for v in (1.6) is therefore the gradient of the tangent to the curve at the instant t; that is, v in (1.6) is the *instantaneous* speed at t. In fact, to make a distinction between this instantaneous speed and the average speed calculated for finite Δt we use the notation of differential calculus* and write:

$$\frac{ds}{dt} = \lim_{\Delta t \to 0} \left\{ \frac{s(t+\Delta t) - s(t)}{\Delta t} \right\}. \tag{1.7}$$

Or, $\dfrac{ds}{dt} = v(t) = 2ct$ the instantaneous speed at t.

Q1 Imagine that the distance travelled by a particle after a time t is given by

$$s = 2 + 3t - t^2.$$

Use equation (1.3) to calculate the average speed in the interval from $t = 2$ to $t = 2 + \delta t$ when $\delta t = 0.1, 0.01,$ and 0.001. What is the instantaneous speed at $t = 2$?

---

*A note on the calculus notation:
The quantity $s(t + \Delta t) - s(t)$ is actually a small increment in s and we might call it Δs. We have avoided writing $\left(\dfrac{s(t+\Delta t) - s(t)}{\Delta t} = \dfrac{\Delta s}{\Delta t} \right)$ because there is tendency for students to fall into the trap of saying that in the limit as $\Delta t \to 0$, $\Delta s \to ds$, and $\Delta t \to dt$. This is of course is nonsense, $\Delta s \to 0$ and $\Delta t \to 0$. In fact, the way in which we have taken the limit in the above analysis is really a mental-scaffolding that allows us to approach the concept of a tangent to a curve in terms of ideas more familiar to our everyday experience such as gradients of chords. Indeed, mathematicians would prefer to treat the symbol d/dt as an 'operation' that one can perform on a function and performing the operation on, say $s(t)$, is effectively asking for the rate of change of $s(t)$ with respect to t; i.e. $(d/dt\,\{s(t)\}) = ds/dt$, which itself is a function of t and gives the instantaneous gradient of s for any t. This operation is called differentiation, and the above procedure followed to calculate the 'derivative' of $s(t)$ is applicable to any function likely to appear in any physics text.

4 Linear mechanics

Q2 The following tables give the distance travelled since $t = 0$. Deduce possible relationships between s and t in each case:

(a)

t	0	1	2	3	4	5
s	0	5	20	45	80	125

(b)

t	0	1	2	3	4	5
s	0	2	4	6	8	10

(c)

t	0	1	2	3	4	5
s	1	4	7	10	13	16

(d)

t	0	1	2	3	4	5
s	0	2	6	12	20	30

1.1.2 The kinematics of falling bodies

Expression (1.6) tells us the instantaneous speed $v(t)$ of a body in free fall. It is of course a motion due to a *uniform acceleration*, say a. In each successive unit of time, the expression (1.6) tells us that the instantaneous speed increases by the value $2c$:

$$v = 2ct. \tag{1.6}$$

Therefore, the *acceleration due to gravity* is $2c$ in units of speed per unit of time – in SI this would be $2c$ m/s per s or $2c$ m/s^2. We usually use the symbol g to represent the acceleration due to gravity, so we see that (1.1) and (1.6) become

$$s = \frac{g}{2}t^2 \quad \text{and} \quad v = gt. \tag{1.8}$$

Hence the odd number progression that emerges out of the free fall of bodies is the signature for a uniform acceleration.

It is informative to apply our method for calculating the instantaneous gradient of a function even though we already know what the answer will be for $v(t)$:

$$\frac{dv}{dt} = \lim_{\Delta t \to 0} \left\{ \frac{v(t + \Delta t) - v(t)}{\Delta t} \right\} = \lim_{\Delta t \to 0} \left\{ \frac{g \cdot (t + \Delta t) - g \cdot t}{\Delta t} \right\} = g. \tag{1.9}$$

This means $\Delta v/\Delta t = g$ whatever the value of Δt, which is of course what one would expect for a uniform acceleration. However, in applying this method we see how the calculus notation naturally arises.

From (1.8)

$$s = \frac{g}{2}t^2, \quad v = gt, \quad a = g,$$

$$\frac{ds}{dt} = gt, \quad \frac{dv}{dt} = g, \quad \frac{d^2s}{dt^2} = g;$$

that is, the gradient function of speed is the acceleration of the body and the gradient function of distance is the speed of the body. The quantity d^2s/dt^2 is called the *second derivative* of s with respect to t and a mathematician would see this as the application of the differentiation operation twice to the function s, hence the notation:

$$\frac{d}{dt}\{v\} = \frac{d}{dt}\left\{\frac{d}{dt}\{s\}\right\} = \frac{d^2s}{dt^2}. \tag{1.10}$$

Section 1.1.3 is a workshop on *differential equations* and looks at a number of the most common $v(t)$, $a(t)$, and $s(t)$ that turn up in elementary kinematics problems.

In Figure 1.3, the body is in free fall so throughout the motion its speed is increasing. Now, it would be simple to compute the distance travelled in the time Δt if the speed of the body were say a constant u throughout the interval. If this were the case, we would simply say

$$\text{Distance travelled} = u \cdot \Delta t. \tag{1.11}$$

This quantity would have the graphical representation of the area shaded in Figure 1.4.

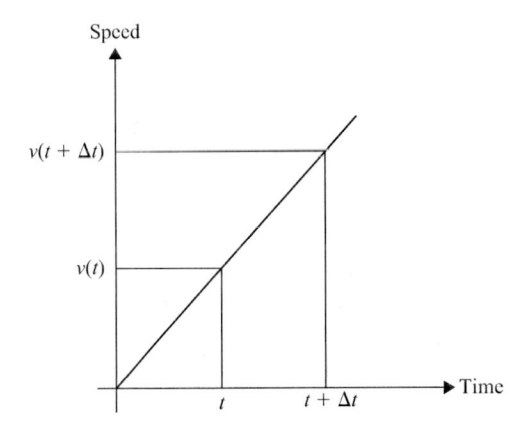

Fig. 1.3

6 Linear mechanics

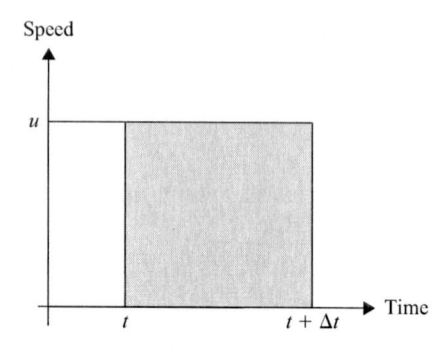

Fig. 1.4

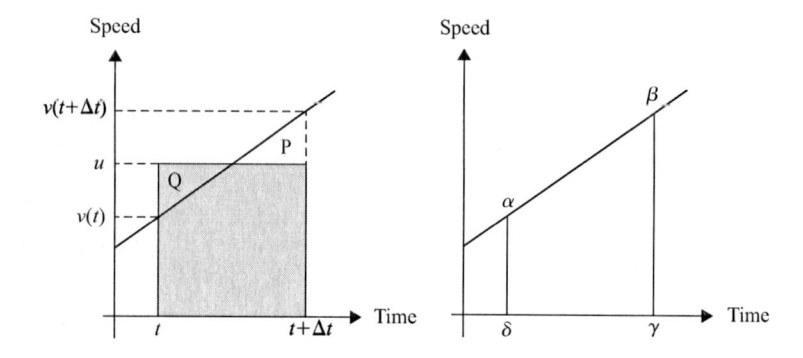

Fig. 1.5

Bearing this in mind we can now see a way of dealing with the question of distance travelled over an interval Δt if the speed is increasing throughout that interval. It is possible to think of the motion described in Figure 1.3 as a motion at a constant speed u where

$$u = \frac{v(t + \Delta t) + v(t)}{2}. \tag{1.12}$$

Effectively we are saying that in the first half of the interval Δt the distance travelled will be less than what it would be if u is given by (1.12), but in the second half of the interval the distance travelled will be more by exactly the distance required to make up the deficit in the first half of the interval. This is more clearly seen in Figure 1.5. We can see by rotating the triangle P into Q that the uniformly accelerated motion over the interval Δt covers the same distance as a motion at a constant speed in (1.12).

Therefore the distance travelled over the interval Δt is

$$s(t + \Delta t) - s(t) = \left(\frac{v(t + \Delta t) + v(t)}{2} \right) \cdot \Delta t, \tag{1.13}$$

which is of course the area of the trapezium α, β, γ, δ.

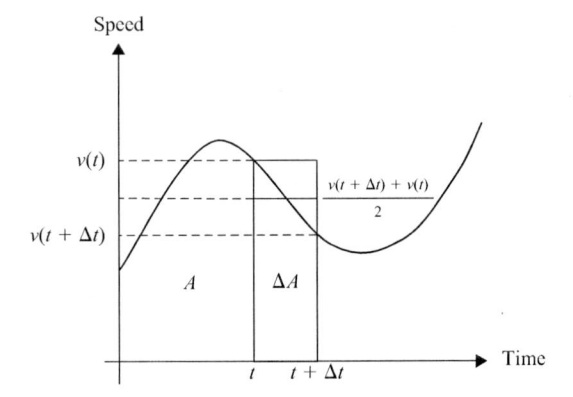

Fig. 1.6

Thus far our exposition has been restricted to the motion of bodies in free fall. However, the methods described above may be used to analyse motions that lead to more complicated functions $s(t)$. In Figure 1.6, the function for speed $v(t)$ is a curve, but this does not stop us from considering the meaning of the area under the curve, even though applying the above methods would only be calculating an approximate value of that area*:

$$\Delta A \approx \left(\frac{v(t+\Delta t) + v(t)}{2}\right) \cdot \Delta t. \tag{1.14}$$

A small increment Δt adds a small area ΔA to the total, which may be approximated by the area of a trapezium.

This of course means that

$$\frac{\Delta A}{\Delta t} \approx \left(\frac{v(t+\Delta t) + v(t)}{2}\right), \tag{1.15}$$

and in the limit as $\Delta t \to 0$

$$\lim_{\Delta t \to 0} \left(\frac{\Delta A}{\Delta t}\right) = \frac{dA}{dt} = v(t); \tag{1.16}$$

that is, the function $v(t)$ is the first derivative of a function $A(t)$. We already know from Section 1.1.2 that the first derivative of $s(t)$ is $v(t)$; however, since the differentiation

---

*This choice of approximation is called 'the midpoint rule' for calculating an approximate value of the area ΔA. Numerical integration (or numerical quadrature) is a whole subject devoted to calculating areas under smooth curves by essentially calculating the areas of very small strips. Many different methods exist, and the midpoint rule is just one of them. In some physics problems, the integration can only be done numerically, so it is important to choose the most appropriate method that leads to a good convergence to an answer for the fewest intervals. When an analytical solution is possible, one is effectively saying that there exists a function that describes in terms of t the area under the speed–time curve (see Figure 1.6) from one t-value to another. Performing the integral is finding this area function.

8 Linear mechanics

operation only calculates the rate of change of a function, $A(t)$ could be

$$A(t) = s(t) + s_o, \qquad (1.17)$$

where s_o is a constant. Differentiating (1.17) would still lead to

$$\frac{dA}{dt} = \frac{ds}{dt} = v(t). \qquad (1.18)$$

This just expresses the fact that the kinematics (the rates of change of quantities) of the motion is unchanged by a shift s_o in the coordinate s. This would correspond to a non-zero value for $s(0)$, or $s(t) = s_o$ at $t = 0$.

The function $A(t)$, and hence $s(t) + s_o$, may therefore be thought of as the limit of the sum of little areas like (1.14) up to a time t:

$$A(t) = \lim_{\Delta t \to 0} \left(\sum_0^t \Delta A \right). \qquad (1.19)$$

Now, just as in differentiation, mathematicians make a distinction between the area calculated for finite Δt (which of course in general is an approximation) and the limiting value of the summation as $\Delta t \to 0$:

$$\lim_{\Delta t \to 0} \left\{ \sum_0^t \left(\frac{v(t + \Delta t) + v(t)}{2} \right) \cdot \Delta t \right\} = \int_0^t v \, dt'. \qquad (1.20)$$

The sign $\int$ is really a very elongated 's' to represent the limiting summation. The term $v \, dt'$ means we are doing this limiting summation to calculate the area under the curve $v(t')$ up to a value of $t' = t$. Therefore t' is often called a *dummy variable* as it is only used to show which variable we are doing the summation over. This gives us a clue as to how we actually calculate such a thing as (1.20).

Since

$$\frac{dA}{dt} = v, \qquad (1.21)$$

then if

$$A(t) = \int_0^t v \, dt', \qquad (1.22)$$

we have $A(t) = s(t) + s(0)$ so

$$s(t) - s(0) = \int_0^t \frac{ds}{dt'} dt'. \qquad (1.23)$$

The quantity on the right is called the *integral* of ds/dt' from $t' = 0$ to $t' = t$. The operation we have performed on ds/dt' is *integration* and we can see why integration is the inverse operation to differentiation.

Of course, all that we have discussed so far for deriving an expression for $s(t)$ by calculating the limiting area under the curve $v(t')$ may be applied to deriving an expression for $v(t')$ by calculating the area under the curve $a(t'')$*; that is,

$$v(t') - v(0) = \int_0^{t'} a\, dt'' = \int_0^{t'} \frac{d^2 s}{dt''^2} dt''. \tag{1.24}$$

Section 7.1 is a workshop on setting up integrals and looks at how integrals occur in physics problems and a general method for setting them up (with a specific example).

Q3 The acceleration a of a particle with speed v along a straight line can have the following forms:
- g
- $-kv$
- $g - kv$,

where g and k are constants. Sketch graphs for each case assuming that $v = 0$ at $t = 0$, of a against v and v against t. +

1.1.3 Workshop: Simple differential equations[†]

You will probably have gleaned much of the calculus appearing in Sections 1.1.1 and 1.1.2 from elementary mathematics courses. However, we have nevertheless included these formal presentations so as to lay down the basics of the calculus language that a physicist must become familiar with.

Very often our analysis of the dynamics of a problem leads us to conclusions about the functions $a(t)$, $v(t)$, and $s(t)$. These conclusions may take the form of explicit functions of t, so straightforward integrations would lead to explicit functions for a, v, and s in terms of t. However, most of the time we do not have this luxury. Take, for example, the case of an object undergoing some v-dependent deceleration. A v-dependence that occasionally turns up in simple dynamics problems is proportional to v, so

$$\frac{dv}{dt} = -kv, \tag{1.25}$$

where k is a constant. The minus sign here is expressing the fact that the acceleration is in the opposite direction to the direction of motion. In Section 1.1.4, we discuss in detail the vector nature of the quantities appearing in (1.25); for now we only

---
*Here t'' is yet another dummy variable to integrate over.
[†]A number of simple differential equations do keep cropping up in elementary physics problems and it is good to be aware of them. Differential equations (a) and (b) at the end of this workshop also turn up in electricity, radioactivity, and many other studies.

10 *Linear mechanics*

concern ourselves with the kinematics problem in one dimension (1-D). This is a *first-order linear differential equation** in v as it gives information about the first derivative of v in terms of v and is not complicated by higher-order powers of the first derivative.

What can we do with (1.25)? Well, it tells us that the function $v(t)$ is such that its gradient at any time t is equal to the function at that time multiplied by a constant, $-k$. What we would like to do is to determine the function of t that has this property. Our discussions in Sections 1.1.1 and 1.1.2 would allow us to give the following approximation:

$$v(t + \Delta t) \approx v(t) + \Delta v = v(t) + \frac{dv}{dt}\Delta t, \qquad (1.26)$$

which gets better as $\Delta t \to 0$. This would mean that $\Delta v = \frac{dv}{dt}\Delta t$, and

$$\Delta v = -kv\Delta t, \qquad (1.27)$$

which rearranges to

$$\frac{1}{v}\Delta v = -k\Delta t. \qquad (1.28)$$

Equation (1.28) is of the form $f(v) \cdot \Delta v = -k\Delta t$, both sides of which can be integrated:

$$\int_{v_o}^{v} f(v')dv' = \int_{v_o}^{v} \frac{dv'}{v'} = -k \int_{0}^{t} dt'.^{\dagger} \qquad (1.29)$$

You should know enough to perform these integrals (certainly you will be able to do the one on the far right). However, the integral $\int_{v_o}^{v} \frac{dv'}{v'}$ leads to the function $\ln(v)$, which you may need some reminding about. Section 7.2 is a workshop on logarithms.

On performing the integrals in (1.29), we get

$$[\ln(v')]_{v_o}^{v} = -k\,[t']_0^t \qquad (1.30)$$

or

$$v(t) = v_o e^{-kt}. \qquad (1.31)$$

With (1.31) we have an explicit function of t for v so it is a simple matter of integration and differentiation to obtain explicit expressions of $a(t)$ and $s(t)$.

---

*Equation (1.25) is in fact second order in x as $v = dx/dt$, so $dv/dt = d^2x/dt^2$. The most important second-order differential equation in x does not appear in Section 1.1.3, but rather has almost two entire chapters devoted to it (Chapters 4 and 5). The differential equation for the *harmonic oscillator* is the basis of all work with waves and oscillations of any kind.

†Notice that the integration variables are primed. As was mentioned in Section 1.1.2, the variables appearing inside the integral sign are dummy variables and represent the 'labels' of the strips being summed. In expression (1.29) of Section 1.1.3, we are integrating the function of v' from an initial value v_o at $t = 0$ to a value v at time t. This of course achieves our goal of finding an expression for v in terms of t.

Obtain explicit expressions of t for $a(t)$, $v(t)$, and $s(t)$ for the systems described by the following differential equations:

(a) $\dfrac{dv}{dt} = -kv$, where k is a constant where $v' = v_0$ at $t' = 0$ and $v' = v$ at $t' = t$

(b) $\dfrac{dv}{dt} = g - kv$, where k and g are constants $v = 0$ when $t = 0$

(c) $\dfrac{dv}{dt} = g - kv^2$, where k and g are constants (*Hint*: partial fractions) and $v = 0$ when $t = 0$

1.1.4 The kinematics of a projectile

Let us analyse what happens when we project a ball straight up into the air. Our hand accelerates the ball to an initial speed, say u, in the upward direction, which the ball has as soon as it leaves the hand. However, once it leaves the hand the ball is in free fall; that is, it begins to accelerate to the ground at a rate g. Now this idea usually takes some getting used to – the ball is moving upwards, but accelerating downwards. To illustrate this more clearly, we usually combine the directional and magnitude information of the above quantities into a single *vector* representation. Thus we talk about the vector **v** as the *velocity* vector and the vector **a** as the *acceleration* vector. In Cartesian coordinates of the x–z plane, one might represent these two vectors using the column matrices

$$\mathbf{v} = \begin{pmatrix} 0 \\ u \end{pmatrix} \tag{1.32}$$

and

$$\mathbf{a} = \begin{pmatrix} 0 \\ -g \end{pmatrix}. \tag{1.33}$$

Notice the negative sign in **a** representing the fact that acceleration is at a rate g in the negative z-direction. The change in the velocity, say $\Delta \mathbf{v}$, of the ball over a time Δt is just

$$\Delta \mathbf{v} = \mathbf{a} \Delta t, \tag{1.34}$$

which means that

$$\mathbf{v}(0 + \Delta t) = \mathbf{v}(0) + \mathbf{a} \Delta t = \begin{pmatrix} 0 \\ u \end{pmatrix} + \begin{pmatrix} 0 \\ -g \end{pmatrix} \cdot \Delta t = \begin{pmatrix} 0 \\ u - g \Delta t \end{pmatrix}. \tag{1.35}$$

The illustration in Figure 1.7 shows how the ball can initially be moving in the upward direction whilst accelerating in the downward direction. The upward velocity becomes reduced by $g \Delta t$ over each interval Δt. Eventually, the velocity is zero. This means that the ball must momentarily be at rest. Then the velocity begins to grow in the negative z-direction until it reaches the velocity $\begin{pmatrix} 0 \\ -u \end{pmatrix}$, at which time it has returned to the point from which it was initially projected.

12 Linear mechanics

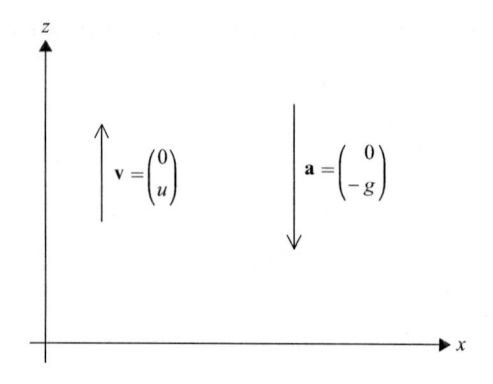

Fig. 1.7

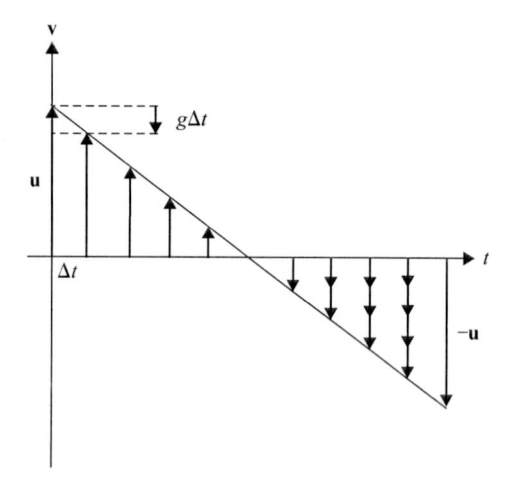

Fig. 1.8

The graph of velocity against time must therefore look like the line drawn in Figure 1.8 The gradient of the line is negative with a magnitude g, which expresses the fact that the acceleration in the z-direction is $-g$. So we may write $\mathbf{v}(t)$ as

$$\mathbf{v}(t) = \mathbf{v}(0) + \mathbf{a}t$$

or

$$\begin{pmatrix} 0 \\ v_z(t) \end{pmatrix} = \begin{pmatrix} 0 \\ u \end{pmatrix} + \begin{pmatrix} 0 \\ -g \end{pmatrix} \cdot t. \tag{1.36}$$

Kinematics 1.1

In this present case of uniform acceleration, we may use the result obtained in (1.13):

$$\mathbf{r}(t) = \frac{1}{2}(\mathbf{v}(0+t) + \mathbf{v}(0))t = \frac{1}{2}\left(\begin{pmatrix} 0 \\ u - gt \end{pmatrix} + \begin{pmatrix} 0 \\ u \end{pmatrix}\right)t = \begin{pmatrix} 0 \\ ut - \frac{1}{2}gt^2 \end{pmatrix}, \quad (1.37)$$

which is a vector called the *displacement* of the ball from the point of initial projection.

If as well as an initial vertical velocity, the ball was simultaneously given an initial horizontal velocity then

$$\mathbf{v}(0) = \begin{pmatrix} u_x \\ u_z \end{pmatrix}, \quad (1.38)$$

where u_x and u_z are the magnitudes of the initial velocities in the x- and z-directions, respectively.

The vector $\mathbf{a}$ would still be given by

$$\mathbf{a} = \begin{pmatrix} 0 \\ -g \end{pmatrix}, \quad (1.39)$$

since there will be no horizontal acceleration once the ball has left the hand.

The vector $\mathbf{r}(t)$ would then be given by

$$\mathbf{r}(t) = \begin{pmatrix} u_x t \\ u_z t - \frac{1}{2}gt^2 \end{pmatrix}. \quad (1.40)$$

In terms of the Cartesian coordinates $x(t)$ and $z(t)$ the vector $\mathbf{r}(t)$ is $\begin{pmatrix} x(t) \\ z(t) \end{pmatrix}$, so we have a pair of parametric equations (in terms of the parameter t):

$$x(t) = u_x t$$

and

$$z(t) = u_z t - \frac{1}{2}gt^2, \quad (1.41)$$

which we can rearrange so that z is a function of x by eliminating t:

$$z(x) = \frac{u_z}{u_x}x - \frac{g}{2u_x^2}x^2. \quad (1.42)$$

Solving the equation $z = 0$ we have two solutions for x:

$$x = 0$$

and

$$x = \frac{2u_x u_z}{g}. \quad (1.43)$$

The first is clearly the origin and therefore the initial point of projection, the second solution is well known to be the range of a projectile projected simultaneously with

14 Linear mechanics

initial velocities u_x and u_z in the horizontal and vertical directions, respectively. The curve in (1.42) is of course a parabola. Notice the time the projectile spends in the air, say τ, is just as it would be if the projectile were given no initial horizontal velocity, that is, $u_x = 0$:

$$\tau = \frac{2u_z}{g}. \tag{1.44}$$

So the range is given by

$$x = u_x \tau = \frac{2u_x u_z}{g}, \tag{1.45}$$

which is just as far as the projectile would cover at a speed u_x over a time τ. Expression (1.42) is not a bad approximation to the kind of curved path that a projectile might really take up if the effect of the air were small. A projectile does not really have separate horizontal and vertical motions. As it moves along its curved path, its motion at any instant is directed along a tangent to the curve. The splitting up of the actual motion along the path into horizontal and vertical motions, called *components*, is a mathematical artifice – the vector representation.

The vector representation therefore seems to give us the correct answer for ideal projectile motion in the absence of air under the influence of gravity. We have still not yet considered the interactions or forces that lead to the acceleration experienced by our projectile (i.e. dynamics and we will deal with that later). However, the validity of the vector representation in this particular case does suggest a law that we often take for granted – that *the horizontal and vertical motions of a body under the influence of gravity alone are independent of each other.*

Q4 A projectile is fired up an inclined plane at some angle to the plane. Let r be the maximum range of the projectile on the plane, and t be the corresponding time of flight. Deduce a relationship between r and t. ++

Q5 A particle is projected from ground level in such a way that it passes through two points at the same height h and a distance d apart. Calculate the direction of the velocity that has the smallest magnitude that just satisfies the above conditions. ++

Q6 A horizontal wind exerts a horizontal force on a projectile of kw, where k is a constant and w is the weight of the projectile. Determine the horizontal range of the projectile and sketch the trajectory for initial horizontal and vertical velocities u_x and u_z. +

1.1.5 Workshop: Motion on the surface of a smooth inclined plane

Suppose a particle is projected with a velocity **v** on the surface of a smooth inclined plane. What path would it take?

We have effectively a projectiles problem again. Let the initial velocity be of magnitude v and let angles A and B both be 30° (see Figure 1.9).

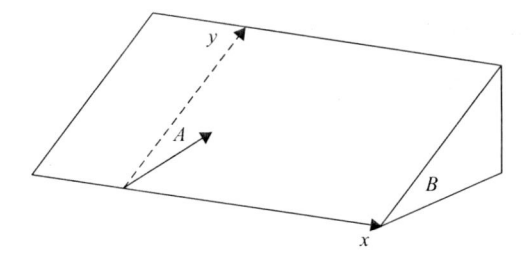

Fig. 1.9

(a) Show that if one chooses coordinate system for the problem as shown above (x, y), then the equation of the path is

$$y = \sqrt{3}x - \frac{gx^2}{v^2}, \tag{1.46}$$

where g is the acceleration due to gravity.

(b) Determine the biggest y-value reached by the particle.

1.1.6 Adding and subtracting vectors

What rules govern resolving, or splitting, a vector into components, and the reverse process of combining components? We saw in the last section the usefulness of a vector representation. However, with the representation comes the rules for combining vectors. To answer these questions we need only look at the addition of vectors. When we represent the displacement $\mathbf{r}(t)$ in the x–z plane by the column matrix $\begin{pmatrix} x(t) \\ z(t) \end{pmatrix}$, we are already assuming the rule for vector addition, that is, we are saying that the displacement $\mathbf{r}(t)$ is equivalent to a displacement $x(t)$ along the horizontal direction and then a displacement $z(t)$ along the vertical direction. It was very convenient to do it this way for the projectile motion because motions in the horizontal and vertical directions were independent of each other and had slightly different rules governing how they changed in time. Adding and subtracting two vectors, like two displacements, follows almost directly from the representation itself. For example, in Figure 1.10 we add and subtract two vectors $\mathbf{r}_1$ and $\mathbf{r}_2$

$$\mathbf{r}_1 + \mathbf{r}_2 = \begin{pmatrix} 4 \\ 3 \\ 5 \end{pmatrix} + \begin{pmatrix} -2 \\ -2 \\ 4 \end{pmatrix} = \begin{pmatrix} 2 \\ 1 \\ 9 \end{pmatrix} ; \quad \mathbf{r}_2 - \mathbf{r}_1 = \begin{pmatrix} -2 \\ -2 \\ 4 \end{pmatrix} - \begin{pmatrix} 4 \\ 3 \\ 5 \end{pmatrix} = \begin{pmatrix} -6 \\ -5 \\ -1 \end{pmatrix}.$$

The result of adding two vectors is to produce a *resultant* $\mathbf{r}_1 + \mathbf{r}_2$. We can see why the resultant is said to be obtained by the *head to tail* rule – the separate addition of x-components, y-components, and z-components is effectively the placing of one vector's *tail* on to the other vector's *head*. Notice it does not matter in which order the addition is done.

16 Linear mechanics

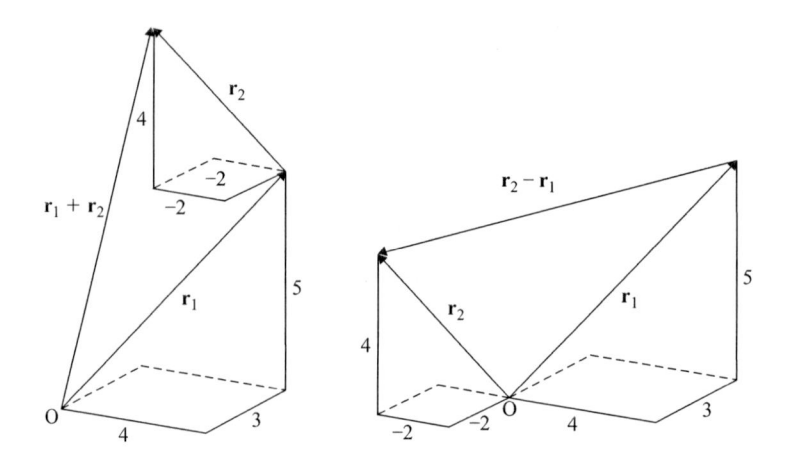

Fig. 1.10

The subtraction $\mathbf{r}_2 - \mathbf{r}_1$ has the meaning of the *relative* displacement; that is, the position of the head of the vector $\mathbf{r}_2$ *relative* to the head of the vector $\mathbf{r}_1$. The relative vector is therefore like shifting the origin of the coordinate system from O to the head of $\mathbf{r}_1$ and viewing the head of $\mathbf{r}_2$ from there. $\mathbf{r}_1 - \mathbf{r}_2$ will of course have the opposite meaning.

It is also convenient to give vectors in terms of unit vectors $\hat{\mathbf{x}}$, $\hat{\mathbf{y}}$, and $\hat{\mathbf{z}}$ defined as follows:

$$\hat{\mathbf{x}} = \begin{pmatrix} 1 \\ 0 \\ 0 \end{pmatrix} \quad \hat{\mathbf{y}} = \begin{pmatrix} 0 \\ 1 \\ 0 \end{pmatrix} \quad \hat{\mathbf{z}} = \begin{pmatrix} 0 \\ 0 \\ 1 \end{pmatrix}.$$

Using these objects, the displacement $\mathbf{r}$ becomes

$$\mathbf{r} = \begin{pmatrix} x \\ y \\ z \end{pmatrix} = x\hat{\mathbf{x}} + y\hat{\mathbf{y}} + z\hat{\mathbf{z}}. \tag{1.47}$$

The rate of change of a vector like $\mathbf{r}$ would then simply be a vector of components which were the rates of change of the components of $\mathbf{r}$. This of course is just the *velocity* vector: $\mathbf{v} = v_x\hat{\mathbf{x}} + v_y\hat{\mathbf{y}} + v_z\hat{\mathbf{z}}$, with $v_x = \dfrac{dx}{dt}$, $v_y = \dfrac{dy}{dt}$, $v_z = \dfrac{dz}{dt}$. Resultant velocities can therefore be calculated in the same way as resultant displacements – head to tail rule applied to velocities.

One would of course obtain relative velocities by subtracting velocity vectors. For example, in Figure 1.10 the vector $\mathbf{r}_2 - \mathbf{r}_1$ could represent the position of a point (2) in a coordinate system with an origin centred on a point (1). Differentiating $\mathbf{r}_2 - \mathbf{r}_1$ would produce the relative velocity of (2) with respect to (1):

$$\frac{d}{dt}(\mathbf{r}_2 - \mathbf{r}_1) = \frac{d}{dt}\mathbf{r}_2 - \frac{d}{dt}\mathbf{r}_1 = \mathbf{v}_2 - \mathbf{v}_1, \tag{1.48}$$

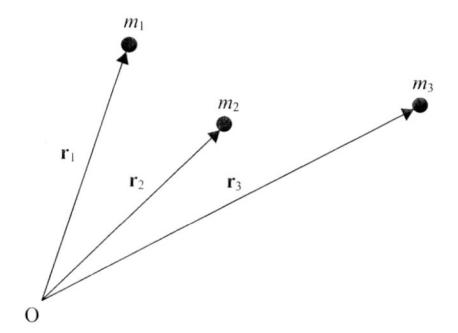

Fig. 1.11

which is also the subtraction of the velocity vector of point (1) from that of point (2) in the coordinate system centred on O. What is the rate of change of a velocity vector **v**? Well, of course the rate of change of **v** is the acceleration vector **a**: $\mathbf{a} = a_x\hat{\mathbf{x}} + a_y\hat{\mathbf{y}} + a_z\hat{\mathbf{z}}$, with $a_x = \dfrac{dv_x}{dt} = \dfrac{d^2x}{dt^2}$, $a_y = \ldots$, $a_z = \ldots$. Addition and subtraction of acceleration vectors produce resultant and relative accelerations, respectively.

Q7 Two aircraft take off from an airfield. One leaves at 14:00 and flies towards east at an average speed of 400 km/h. The other leaves at 14:10 and flies on a course 60° east of north at an average speed of 450 km/h. Find their distance apart at 15:00 and the bearing of the first with respect to the second.

Q8 A particle describes a circle of radius a with a uniform speed v. If it starts from a point X, determine an expression for the displacement from X at a time t.

Q9 Rain falls vertically at 8.0 m/s. The rain drops make tracks on the side of a car window at an angle 30° below the horizontal. Calculate the speed of the car.

Q10 Resolve a velocity of 10 m/s into two perpendicular components such that
- the components are equal
- one component is twice the other.

1.2 Dynamics

1.2.1 Newton's laws

Newton's laws of motion give no specific details about interactions that exist between bodies in the universe, but they do have much to say about the consequences of these interactions. In Figure 1.11, we have three interacting masses m_1, m_2, and m_3 far removed from any other bodies so that the three masses interact with each other and with nothing else.

From more elementary work, we remind ourselves that the momentum of a body is defined as the *mass* of the body multiplied by its *velocity*:

$$\mathbf{p} = m\mathbf{v}. \tag{1.49}$$

18 Linear mechanics

The momentum is used to describe *the state of motion* of a body, and Newton equates the rate of change of momentum to the resultant force applied.

$$\mathbf{F} = \frac{d}{dt}\mathbf{p}. \tag{1.50}$$

Thus the state of motion of a body can only change if there is a resultant force applied.

The momenta of the three masses in Figure 1.11 are

$$\mathbf{p}_1 = m_1\frac{d}{dt}\mathbf{r}_1 = m_1\mathbf{v}_1, \quad \mathbf{p}_2 = m_2\frac{d}{dt}\mathbf{r}_2 = m_2\mathbf{v}_2, \quad \mathbf{p}_3 = m_3\frac{d}{dt}\mathbf{r}_3 = m_3\mathbf{v}_3.$$

The force accelerating m_1 is

$$\mathbf{F}_{12} + \mathbf{F}_{13} = \frac{d}{dt}\mathbf{p}_1 = \frac{d}{dt}(m_1\mathbf{v}_1), \tag{1.51}$$

where $\mathbf{F}_{12}$ and $\mathbf{F}_{13}$ are the forces exerted on m_1 by m_2 and m_3, respectively.

Similar expressions for masses m_2 and m_3 exist:

$$\mathbf{F}_{21} + \mathbf{F}_{23} = \frac{d}{dt}\mathbf{p}_2 = \frac{d}{dt}(m_2\mathbf{v}_2). \tag{1.52}$$

$$\mathbf{F}_{31} + \mathbf{F}_{32} = \frac{d}{dt}\mathbf{p}_3 = \frac{d}{dt}(m_3\mathbf{v}_3). \tag{1.53}$$

If there are no interactions with anything outside the system, these forces must necessarily sum to zero as not doing so would imply a non-zero resultant force on the system as a whole. So,

$$\mathbf{F}_{12} + \mathbf{F}_{13} + \mathbf{F}_{21} + \mathbf{F}_{23} + \mathbf{F}_{31} + \mathbf{F}_{32} = \mathbf{0}, \tag{1.54}$$

which must of course give us:

$$\mathbf{0} = \frac{d}{dt}(\mathbf{p}_1 + \mathbf{p}_2 + \mathbf{p}_3) = \frac{d}{dt}(m_1\mathbf{v}_1 + m_2\mathbf{v}_2 + m_3\mathbf{v}_3); \tag{1.55}$$

that is,

$$m_1\mathbf{v}_1 + m_2\mathbf{v}_2 + m_3\mathbf{v}_3 = \text{constant} = \mathbf{P}. \tag{1.56}$$

So no matter how these bodies move, the vector sum of the three momenta $\mathbf{p}_1$, $\mathbf{p}_2$, and $\mathbf{p}_3$ must always equal a constant vector $\mathbf{P}$ for all time (see Figure 1.12).

Now,

$$m_1\mathbf{v}_1 + m_2\mathbf{v}_2 + m_3\mathbf{v}_3 = \frac{d}{dt}(m_1\mathbf{r}_1 + m_2\mathbf{r}_2 + m_3\mathbf{r}_3) = \mathbf{P}, \tag{1.57}$$

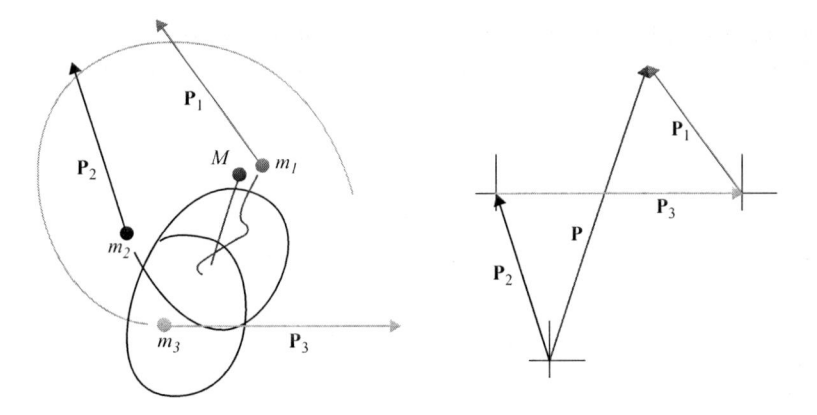

Fig. 1.12

as m_1, m_2, and m_3 are all constants. So integrating (1.57) gives us

$$m_1\mathbf{r}_1 + m_2\mathbf{r}_2 + m_3\mathbf{r}_3 = \mathbf{P}t + \mathbf{Q}, \tag{1.58}$$

where $\mathbf{Q}$ is some constant vector.

The left-hand side of (1.58) may now look familiar. If we were looking for the coordinates of the *centre of mass** of the system of three masses, we would write

$$m_1\mathbf{r}_1 + m_2\mathbf{r}_2 + m_3\mathbf{r}_3 = M\mathbf{R}, \tag{1.59}$$

where $M = (m_1+m_2+m_3)$, the total mass of the system, and $\mathbf{R}$ is the position vector for the centre of mass of the three masses. This means that

$$\mathbf{R} = \frac{1}{M}(\mathbf{P}t + \mathbf{Q}). \tag{1.60}$$

This just expresses the fact that the centre of mass drifts at constant velocity (say, $\mathbf{V}$) whilst the three masses move in orbit around each other. The point in Figure 1.12[†] labelled M drifts as the masses m_1, m_2, and m_3 orbit around each other (notice that the path of M is parallel to the constant vector $\mathbf{P}$). $\mathbf{P}$ turns out therefore to be not only the total momentum of the system but also the momentum of the centre of mass; that is,

$$\mathbf{P} = \frac{d}{dt}(M\mathbf{R}) = M\frac{d}{dt}\mathbf{R} = M\mathbf{V}. \tag{1.61}$$

By starting with equations (1.55) and (1.57) and integrating, we have introduced constant vectors $\mathbf{P}$ and $\mathbf{Q}$. We can see though that with an appropriate choice of our

---
*Section 7.7 is a workshop on locating centres of mass.
[†]Figure 1.12 was created using a computer model and rendering the output as an animation. The figure is in fact a screen shot of the animation.

20   Linear mechanics

point of view, we could make both of these constants zero and this would have no effect on the dynamics of the system. Different points of view appear in the general expressions and merely represent a difference in uniform relative motion of the coordinate system chosen to describe the motion of the three masses.

The vector sum of $\mathbf{p}_1$, $\mathbf{p}_2$, and $\mathbf{p}_3$ must always equal $\mathbf{P}$ – the centre of mass just drifts along at a constant velocity because there is no resultant force from outside the system to accelerate it. This is of course traditionally called Newton's first law, whilst equating the rate of change of momentum to the resultant force is the second law. The constant state of motion represented by the constant vector $\mathbf{P}$ is essentially a statement of the law of conservation of linear momentum as well as a statement of the first law. The so-called third law* ensures that bodies in the universe do not just act on each other – they interact; that is, the internal forces that occur within our closed system are such that a vector sum of them must always be zero.

Let us summarize all this in four statements:

- Bodies continue in their state of motion (represented by a momentum vector) unless a resultant force acts upon them.
- The resultant force on a body is equal to the rate of change of the momentum vector of the body.
- Forces arising directly from an interaction between two bodies must always sum to zero.
- The sum of linear momenta in a closed system† is a constant of the motion.

Q11 Six identical numbered cubes, each of mass m, lie in a straight line on a smooth horizontal table. A constant force $\mathbf{F}$ is applied along the line of cubes. Determine
  (a) the acceleration of the system
  (b) the resultant force on each cube
  (c) the force exerted on the fifth cube by the fourth cube. +

Q12 A bullet of mass 20 g is shot into a block of wood with an initial speed of 300 m/s, and is brought to rest in 0.01 s. Calculate the resistive force exerted by the wood assuming it to be constant over the 0.01 s.

---

*In more recent times, the use of the word *reaction* in the statement of Newton's third law has caused dreadful confusion with students. In applied mathematics, the word reaction has come to mean *normal reaction* at a surface of contact. An object sitting on the surface of the Earth would not accelerate towards the centre of the Earth because of the normal reaction from the ground. However, the '*equal and opposite reaction force*' of Newton's third law for the gravitational interaction between the object and the Earth is actually the force of attraction (to the object) acting at the centre of the Earth and NOT the normal reaction from the ground. To see this, just imagine lifting the object up and dropping it – Newton's third law still applies, but with no normal reaction the object and Earth accelerate towards each other (of course the Earth moves an infinitesimally small distance).

†The system in our example of Figure 1.12 is what physicists call closed because there is no interaction between the masses in the system and anything outside the system (since there is nothing at all outside the system). In reality it is impossible to isolate such systems from the rest of the universe so the law of conservation of linear momentum is always approximately upheld and agreement is close when the interactions between the system in question and the rest of the universe can be neglected.

Q13 A man of mass 70 kg stands on a set of bathroom scales on the floor of a lift. Find the reading registered by the scales when
  (a) the lift is ascending with a uniform acceleration of 5 m/s^2
  (b) the lift is descending with a uniform acceleration of 5 m/s^2.

  Assume that the scales were calibrated in a gravitational field of strength 9.81 N/kg. +

1.2.2 The principle of relativity

Let us now consider implications with a much simpler example. Imagine two spacecraft of equal mass, approaching each other for docking in deep intergalactic space so that interactions with anything else may be said to be negligible.

Imagine monitoring the event from the centre of mass coordinate system. The two craft approach each other at equal and opposite velocities $+\mathbf{v}$ and $-\mathbf{v}$ (their relative velocity must be $2\mathbf{v}$) and dock at the origin of the centre of mass coordinates.

From the point of view of the centre of mass coordinate system, the whole docking procedure is symmetrical – there is no reason for the two docked craft to accelerate in any particular direction. Actually Newton's laws allow you to predict more than this. Even if the docking were to misfire and the two craft were to rebound off each other, whatever their rebound velocities, the two velocities would have to be equal and opposite (the momentum vectors must, in the centre of mass coordinate system, sum to zero).

Now consider the event from the point of view of a coordinate system attached to one of the craft. You would see the other craft approaching at velocity $2\mathbf{v}$, but on impact you would see both the craft move in the direction of the other craft at a reduced velocity $\mathbf{v}$. You are now in a coordinate system moving at velocity $\mathbf{v}$ with respect to the centre of mass frame, so the docked craft now appear to be moving at velocity $\mathbf{v}$. The difference between the two events occurs because of a difference in uniform relative motion of the two coordinate systems chosen to describe them. The two events are the same, so the physics (the laws of motion) in one must be the same as the physics in the other. Put another way: *all points of view moving with different uniform velocities are as good as one another for the description of the physics*; that is, the constants of integration $\mathbf{P}$ and $\mathbf{Q}$ in our example are merely informing us that the calculus we use to encode the kinematics is insensitive to the uniform relative motion of different points of view. Coordinate systems that share this property for the description of physics are called *Galilean*, after Galileo who was one of the first to realize this remarkable property of nature – the physics of an event cannot be different just because you walk past it. This is the *principle of relativity*. The extension of the principle of relativity to include electromagnetism is what physicists call the special theory of relativity.

In Figure 1.12, our system of three interacting masses seems to drift past at the velocity of the centre of mass because the coordinate system chosen to analyse the system was chosen with a uniform relative motion with respect to the centre of mass of the system.

1.2.3 Impulse and impulsive forces

In a closed system any change in momentum of one part of the system must be balanced by an equal and opposite change in momentum of the rest of the system, otherwise the total momentum would not be conserved and this would imply the action of an external influence (which is absurd since the system is supposed to be closed).

The term *impulse* of a force is defined as follows:

When a force **F** is constant, the impulse is the product of the force and the time during which it acts, that is **F**Δt. Note that it is a vector quantity. If the force acts on a constant mass m, it produces an acceleration **a** so that

$$\mathbf{F}\Delta t = m\,\mathbf{a}\,\Delta t = m(\mathbf{v} - \mathbf{u}), \tag{1.62}$$

which is the change in momentum of the mass m.

When **F** is not a constant force, the impulse of **F** becomes

$$\int_0^t \mathbf{F}\,dt', \tag{1.63}$$

where t is the time during which the force acts and t' is the dummy variable of integration. **F** is equal to the rate of change of momentum of the mass m:

$$\mathbf{F} = \frac{d}{dt}\mathbf{p}. \tag{1.64}$$

So,

$$\int_0^t \mathbf{F}\,dt' = \int_0^t \frac{d}{dt'}\mathbf{p}\,dt' = \mathbf{p}(t) - \mathbf{p}(0), \tag{1.65}$$

which is once again the change of momentum produced by the force applied over that time. The impulse of a force applied over a time t is thus the change in momentum produced.

Occasionally, in problems the applied force **F** may be very large, but acting over a short time. The body would only move a very short distance while the force is acting. In the limiting situation of an infinitely great **F** acting over an infinitely short time, the change in position 'during' the interaction would be zero. This limiting situation is known as an *impulsive force*. Equation (1.65) would still hold in this case.

In Figure 1.13, the two idealized forces both impart the same impulse to a body that they act upon:

$$\mathbf{F}_1\Delta t_1 = \mathbf{F}_2\Delta t_2, \tag{1.66}$$

however, $\mathbf{F}_1$ is a small force applied over a long time whereas $\mathbf{F}_2$ is a larger force applied over a shorter time. The area under such a force – time graph is the impulse and in the limit* of an impulsive force $\Delta t \to 0$ as $\mathbf{F} \to \infty$. This limit is of course never realized

---

*As $\Delta t \to 0$ and $\mathbf{F} \to \infty$, we reach an idealized limit where the area $\int \mathbf{F}\,dt'$ remains finite as it is the impulse or change in momentum. The object that is the limit of this process is called the delta distribution and it is usually given by the symbol $\delta(t)$. Such objects become very useful in more advanced work.

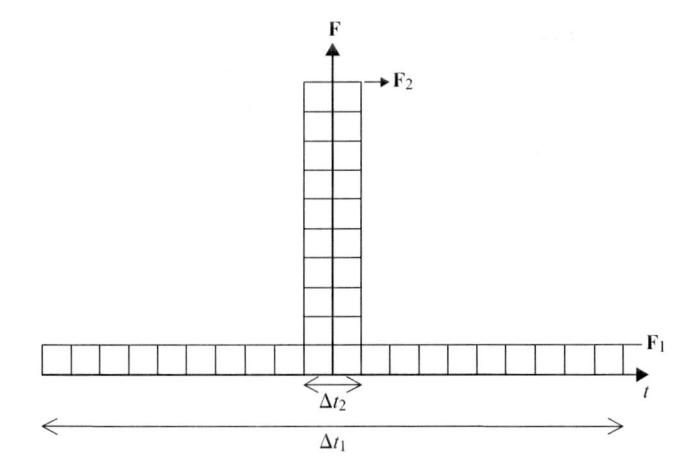

Fig. 1.13

in practice, but approximate examples are the blow of a hammer or the collision of hard billiard balls.

Q14 A hot air balloon ascends at a steady speed of 10 m/s. The pilot drops a 10 kg sandbag from a height of 600 m. Calculate the magnitude of the impulse with which the sandbag hits the ground. Assume no rebound at the ground.

Q15 A shell of mass m is fired from a gun of mass M placed on a smooth horizontal track. The barrel of the gun is inclined at an angle θ to the horizontal and imparts the recoil motion of the gun to the shell before the shell finally leaves the barrel. Show that the initial direction of the shell's motion is inclined at an angle α:

$$\alpha = \tan^{-1}\left[\left(1 + \frac{m}{M}\right)\tan\theta\right]. ++$$

Q16 A particle of mass m is moving at 5 m/s in a given direction and is struck by an impulsive force **F** that deflects its direction of motion through 60° and doubles the magnitude of its velocity. The same impulsive force **F** is applied to a mass of $5m$ at rest. Describe the resulting motion of the $5m$ mass.

1.2.4 Workshop: The conservation of linear momentum

As an example of the use of the conservation of linear momentum in problems, we shall consider a two-dimensional (2-D) collision between two hard spheres. Here we are considering smooth spheres, so the mutual action between them is then along the line joining their centres. Therefore, the state of motion perpendicular to the line joining their centres remains unchanged. When hard spheres collide they are slightly compressed and return to their original shape causing a rebound. The collision can therefore be divided into two parts: impact and compression, and restitution and separation. The property that causes bodies to recover their original shape is called

24 Linear mechanics

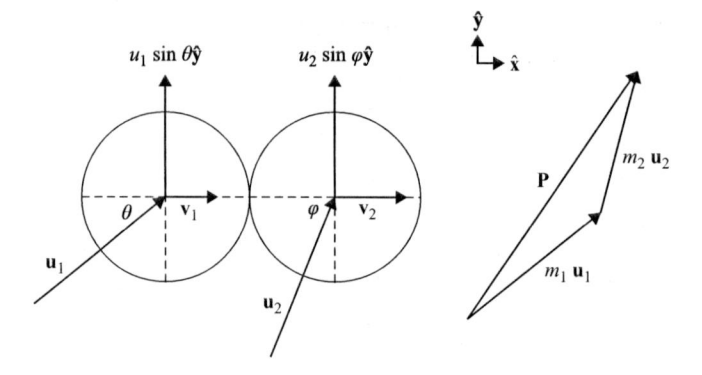

Fig. 1.14

elasticity. However, this property depends on the material of which the bodies are made; whilst the law of conservation of linear momentum always holds true, there is no way of calculating the effect of the elasticity of the bodies* in the collision. We therefore have to fall back on the results of experiments first conducted by Newton himself. Newton discovered the following empirical law:

When two bodies of given substances collide the relative velocity after impact is in a constant ratio to the relative velocity before impact, and in the opposite direction. If the bodies impinge obliquely the empirical law holds for the component velocities along the common normal.

In Figure 1.14 two smooth hard spheres of mass m_1 and m_2 impact obliquely. We have chosen our coordinate system so that the spheres remain on the x–y plane and the line through their centres is parallel to the x-axis. The initial and final velocities of the spheres are

$$\mathbf{u}_1 = \begin{pmatrix} u_1 \cos\theta \\ u_1 \sin\theta \end{pmatrix} \quad \mathbf{u}_2 = \begin{pmatrix} u_2 \cos\phi \\ u_2 \sin\phi \end{pmatrix}$$

$$\mathbf{v}_1 = \begin{pmatrix} v_1 \\ u_1 \sin\theta \end{pmatrix} \quad \mathbf{v}_2 = \begin{pmatrix} v_2 \\ u_2 \sin\phi \end{pmatrix}.$$

With nothing else involved in the collision, the momentum vector diagram shows how the initial momenta $\mathbf{p}_{i1}$ and $\mathbf{p}_{i2}$ (and indeed the final momenta, $\mathbf{p}_{f1}$ and $\mathbf{p}_{f2}$) must sum to a constant vector $\mathbf{P}$:

$$\mathbf{p}_{i1} + \mathbf{p}_{i2} = \mathbf{p}_{f1} + \mathbf{p}_{f2} = \mathbf{P}$$

$$\mathbf{p}_{i1} = m_1 \mathbf{u}_1$$

$$\mathbf{p}_{i2} = m_2 \mathbf{u}_2$$

---

*To work out the effect of the elasticity, one would have to know all about the internal structure of the colliding bodies. Newton's empirical law conveniently encompasses the effect of the motions and interactions of the internal parts of the colliding bodies in terms of the relative velocities of the bodies before and after impact.

$$\mathbf{p}_{f1} = m_1 \mathbf{v}_1$$
$$\mathbf{p}_{f2} = m_2 \mathbf{v}_2.$$

(a) Use Newton's experimental law to show that
$$\frac{v_1 - v_2}{u_2 \cos \phi - u_1 \cos \theta} = e,$$
where e is a constant.

(b) Obtain the following expressions for v_1 and v_2:
$$v_1 = \frac{(m_1 - em_2)u_1 \cos \theta + m_2 u_2 (1+e) \cos \phi}{m_1 + m_2}$$
$$v_2 = \frac{(m_2 - em_1)u_2 \cos \phi + m_1 u_1 (1+e) \cos \theta}{m_1 + m_2}.$$

(c) If $m_1 = m_2$ and $\mathbf{u}_2 = \mathbf{0}$ (the second sphere is initially at rest), then show that
$$\frac{v_1}{v_2} = \frac{(1-e)}{(1+e)}.$$

(d) What does the expression in (a) imply when $e = 1$?

Q17 Two identical smooth elastic spheres ($e = 1$), each of mass m, are at rest on a smooth horizontal table and touch each other. They are struck symmetrically by an identical sphere, but of mass M, having a velocity $\mathbf{u}$ perpendicular to the line of centres of the two stationary spheres. The mass M comes to rest immediately after the collision. Determine the ratio M/m.

Q18 Three identical particles X, Y, and Z of mass m are placed on a smooth horizontal table. X is joined to Y and Z by light (can be thought of as massless) inextensible strings XY and XZ. The angle XYZ is 60°. An impulse I is applied to X in the direction YX. The strings act as constraints so that the initial motions of Y and Z must be the same as the components of the initial motion of X along YX and ZX, respectively. Determine the initial velocities of the particles. +

Q19 If the initial relative velocity of two spheres before direct impact has a magnitude u, show that the magnitude of the impulse each sphere receives is given by
$$\left(\frac{(1+e)mM}{m+M} \right) u,$$
where m and M are the masses of the spheres. +

Q20 Three spheres of equal mass and coefficient of restitution e are initially at rest in a straight line. The first sphere is given an initial speed u. Determine the speeds of the three spheres after two impacts. +

1.2.5 The law of falling bodies

In Section 1.1, we introduced the law of falling bodies. Within the framework of Newton's laws of motion, a little more insight may be had. Newton's second law

26 Linear mechanics

applied to a body of constant mass m acted upon by a resultant force $\mathbf{F}$ gives

$$\mathbf{F} = \frac{d}{dt}\mathbf{p} = \frac{d}{dt}(m\mathbf{v}) = m\frac{d}{dt}\mathbf{v} + \mathbf{v}\frac{d}{dt}m = m\frac{d}{dt}\mathbf{v} = m\mathbf{a}, \qquad (1.67)$$

where $\mathbf{v}$ is the instantaneous velocity of the body and $\mathbf{a}$ is the acceleration of the body, as $\frac{d}{dt}m = 0$ because m is a constant. The mass m appearing in expression (1.67) essentially describes how hard it is to accelerate the body and is for this reason called the *inertial mass* of the body.

Now, if the accelerating force arises from a gravitational interaction between the body and say the Earth, the accelerating force would be the weight $\mathbf{W}$ of the body:

$$\mathbf{W} = m\mathbf{g}, \qquad (1.68)$$

where $\mathbf{g}$ is the *gravitational field strength* vector. The mass m appearing in this expression (1.68) determines how hard the Earth will pull on it, and by Newton's third law this m also determines how strongly the body will attract another body, the Earth. We therefore call this mass the *gravitational mass*.

The two letters m appearing in both (1.67) and (1.68) do not seem, *a priori*, to be representing the same kinds of things. Newton realized though that the law of falling bodies implies that they must at least be related by a direct proportion – if we double the mass of the body in a gravitational field we would double the weight and hence the accelerating force in free fall, but we would also double how hard it is to accelerate the body, thus *the motion of free fall is universally the same, independent of the size and material of the body.**

Newton used these ideas to further develop his thoughts on gravity. The apocryphal story of Newton and the apple may well be fictitious, but it illustrates very beautifully with the aid of reasonable common experience one of the most profound discoveries ever made. Newton is supposed to have compared the fall of an apple near the surface of the Earth and the motion of the Moon around the Earth. In seeking to apply the law of falling bodies to both the Moon and the apple, Newton demonstrates the universality of his laws of motion.

1.2.6 Workshop: Newton and the apple

In this workshop, you will follow the thoughts that convinced Sir Isaac Newton of his universal law of gravitation. Near the surface of the Earth, an apple, when released from rest, accelerates towards the centre of the Earth at 9.81 m/s^2. Newton knew (from the Greeks) that the Moon is 60 times further from the centre of the Earth than an apple at the Earth's surface. He also expected the acceleration due to gravity at

---

*This is a surprising property of nature, though after some thought one might think it entirely obvious – both types of mass go by the amount of matter present. For Einstein though this property was too intriguing just to leave it at that.. If no experiment can distinguish between these two kinds of mass, then does this mean that our framework of nature must be such that we *cannot* tell them apart? Einstein's general theory of relativity, which is beyond the scope of this book, essentially describes a framework in which gravitational and inertial mass are necessarily the same.

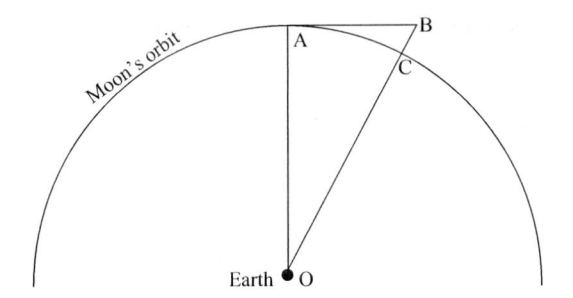

Fig. 1.15

the Moon's orbit to be smaller than at the surface of the Earth. The question was, how much smaller? Sixty times?

Newton had a remarkable intuition, and expected the acceleration due to gravity not to be 60 times smaller, but 60 × 60 times smaller, that is, 3600 times smaller.

(a) Show that the Moon must fall about 1.3 mm in 1.0 s.

The Moon takes 28 days to orbit the Earth, at a radius of 384 000 km.

(b) Show that the speed of the Moon in its orbit is approximately 1.0 km/s.

Study Figure 1.15, which is *not to scale*.

Newton argued (by evoking his first law) that if there were no gravitational acceleration towards the centre of the Earth, the Moon would continue to move along the line AB. Therefore the Moon travelling in its orbit actually falls below the line AB towards the centre of the Earth.

In (a), you showed that each second, the Moon must fall about 1.3 mm.

(c) Show that a fall of about 1.3 mm each second is just what is required to describe the motion of the Moon in its circular orbit.* (*Hint*: Consider the triangle OAB and determine an expression for the distance BC in terms of the other parameters; do not forget BC will be small compared to the other parameters if it is the distance the Moon falls each second.)

1.3 Conclusion

In Section 1.1, we saw how the motions of particles could be described within a framework of 3-D space realized in the course of time. The functions $s(t)$, $v(t)$, and $a(t)$, with the help of vector notations, completely describe the trajectory of a particle. Calculus is the language of translation between these functions and much can be gleaned about the motion through $s(t)$, $v(t)$, and $a(t)$ without actually appealing to the cause or dynamics of the motion at all.

---

*Having demonstrated this astounding result, Newton then left this calculation for nearly 10 years before he picked it up again. It seems that he preoccupied himself in demonstrating that the whole Earth acted on apples and the Moon as though all of the Earth's mass were concentrated at its centre. For more on this and other similar problems, please take a look at the Gauss' law in Chapter 2.

The description of the motion of a particle involves the rates of change of functions, so it is not a surprise that constants of integration arise out of initial conditions. However, we saw in Sections 1.2.1 and 1.2.2 that physics seems insensitive to the uniform relative motion of points of view and therefore *all points of view moving with different uniform velocities are as good as one another for the description of the physics.*

If it is possible to treat a system of particles as closed, we can rely on the idea that the centre of mass of the system can only be at most in uniform relative motion with respect to our point of view. The total momentum of the system of particles must then be equal to, for all time, some constant vector. This total momentum turns out to be the momentum of the centre of mass. A clever choice of coordinate system can get rid of this relative motion, and then the total momentum of the system would be zero, or put another way, our point of view would then be attached to the centre of mass.

2
Fields

Our previous chapter was in two parts. The first part showed you how an acceleration would affect an object. The second part explained that forces could cause accelerations. A question was left unanswered, 'What causes the forces?'. You may have also found it strange that one vital concept was completely missing from the whole chapter – nowhere was energy mentioned. It is time to set both of these deficiencies right, and the topic of fields is the best place in which to do it.

Physics is the science of matter and interactions, and the latter are more important, since we could not understand the structure of the atom, say, without some idea of the forces which hold it together. And by far the most helpful way of viewing interactions is in terms of fields. And when objects interact, there is a certain something which can be passed from one object to another via interactions with fields – a something which is passed on but never created or used up – a something which helps us understand the very nature of interactions – a something which also makes our calculations much easier. That something is, of course, energy.

2.1 Introduction and field strength

Every school student knows that a positive charge will attract a negative charge and repel another positive one. Many also know the formula for the magnitude of the repulsion: $F = Qq/4\pi\varepsilon d^2$, where a negative value of F implies attraction. It is often helpful to view this force as a two-stage process. Rather than saying that Q repels q directly, we say that Q does something to the surrounding space. When q is put in this space it experiences a force, because of what has been done to its location by the big charge Q. This 'affecting of space' is called a field.

A point charge Q sets up an electrostatic field of magnitude $E = Q/4\pi\varepsilon d^2$ at all points around it, where d is the distance of that point from the charge Q. When a second charge q is put into this field, it experiences a force of magnitude $F = qE$. The advantage of this way of thinking is that whatever kind of charge q we are putting into the field, the first part of the calculation (working out E) is always the same.

We may summarize by saying that for any interaction, there are two relationships we need to find. First we need to calculate the field strength at each point in space for a given arrangement of charges, masses, currents, and nuclear dipole moments (or whatever we have in the system). Second we need a formula which tells us how the field at a given point will affect an object placed there.*

---

*In more complicated electrostatic work, we use a field within a field. We say that the system of charges sets up a D-field, which in turn interacts with the material in that region to form the E-field

30 Fields

For electric fields, we define the field strength at a point $\mathbf{E}$ as the force per unit charge which would act on a small test charge placed there. Accordingly, $\mathbf{E}$ is a vector, and, provided that our test charge does not disturb the arrangement of the other charges in the system, the force on a point charge q placed at a point $\mathbf{r}$ is given by $\mathbf{F}(\mathbf{r}) = q\mathbf{E}(\mathbf{r})$.

Similarly, we define the gravitational field strength $\mathbf{g}$ at a point as the force per unit mass which would act on a small test mass there. Accordingly, $\mathbf{g}$ is a vector, and, provided that our test mass does not disturb the arrangement of the other masses in the system, the force on a point mass m placed at a point $\mathbf{r}$ is given by $\mathbf{F}(\mathbf{r}) = m\mathbf{g}(\mathbf{r})$.

As indicated by the bold print, one vital fact about our gravitational and electrostatic fields is that they are vector fields because the field strength $\mathbf{E}$ or $\mathbf{g}$ has directionality as well as magnitude.

Q1 Two positive charges are fixed on the x-axis. One, at the origin, has a charge Q, the other is at $x = 0.5$ m and has charge q. Where can I put a small test charge so that it will not accelerate? If I move the small charge a very small distance from its equilibrium position, will the resultant force on it tend to push it back to equilibrium? +

Q2 The gravitational field strengths in London, Paris, and Chennai are 9.81183, 9.80943, and 9.78281 N/kg, respectively. Why are they different? If I took scales made in London to Chennai, and there measured my weight as '72.00 kg', what would my real mass be?

Q3 If Earth and Sun were given positive electric charges in proportion to their masses, how much charge would the Earth have to have before the electrostatic repulsion was equal and opposite to the gravitational attraction?

2.2 Workshop: Motion in a uniform field in one dimension

We start with the simplest kind of situation. An object is going to move subject to only one force – gravity. At the surface of the Earth, we can approximate the gravitational field strength as constant and uniform, and we call it g. The force on the object (its weight) is therefore $W = mg$, and this is also constant. Let us allow the object to fall from height h above the ground until it hits the ground. It will accelerate downwards with acceleration $a = W/m = g$ as shown in figure 2.1.

In Section 1.1, you analysed this kind of motion in terms of the time elapsed since it was released. Here we shall do an analysis in terms of the distance fallen.

---

which actually causes the force on the test charge. Gauss' law of electrostatics allows us to calculate the D-field at the location desired, we then use knowledge of the polarizability of the material to obtain a relationship linking the D-field with the E-field this will make at this point. For a single point charge Q, the D-field at a distance d from the charge Q is given by $D = Q/4\pi d^2$. In a linear, isotropic, dielectric material, the electric field is given by $E = D/\varepsilon_0 \varepsilon_{\text{rel}}$, and the force on a test charge q is $F = qE = Qq/4\pi\varepsilon_0\varepsilon_{\text{rel}}d^2$, as we would expect by Coulomb's law. More complicated materials have a more complex relationship between D and E, but this increased complexity need not obscure the clarity of the calculation of the D-field in the first place.

Workshop: Motion in a uniform field in one dimension 2.2

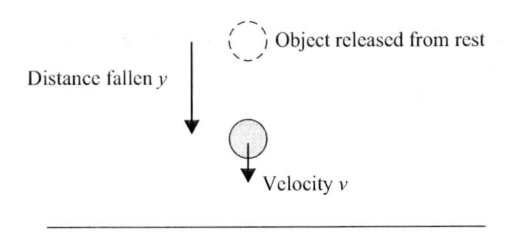

Fig. 2.1

Object released from rest, Distance fallen y, Velocity v

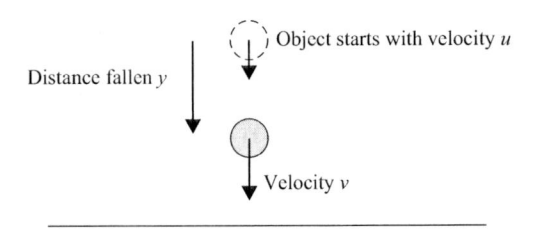

Fig. 2.2

Object starts with velocity u, Distance fallen y, Velocity v

(a) If we use the function $y(t)$ to represent the distance, the object has fallen at time t. Our earlier studies (in section 1.1.2) show us that the velocity $v(t)$ is given by $v = at = gt$, while $y(t) = y = \frac{1}{2}gt^2$. Combine these equations to find v as a function of y and g (i.e. eliminate t from the equations).

(b) We are now going to derive the same result using a more qualitative method. For an object in uniform acceleration, its average velocity* is equal to the arithmetic mean of its initial and final velocities. Show that this enables us to write $y = \frac{1}{2}vt$ for our object dropped from rest until (at time t later) it has velocity v. Given that $v = gt$, eliminate t from the equations to give v as a function of y and g. Check it agrees with your answer to (a).

(c) So far, we have let the object start from rest. We shall now take its initial downward velocity as u, as illustrated in Figure 2.2. Repeat the methods of either (a) or (b) to find an equation without t linking the distance an object falls y to the acceleration g, the initial downward velocity u and the final downward velocity v.[†]

(d) Show that the product of the force (mg) with the distance fallen is equal to $\frac{1}{2}mv^2 - \frac{1}{2}mu^2$.

---

*The definition of average velocity is, of course, distance travelled ÷ time taken, or equivalently it is the steady velocity which would have enabled the object to cover that distance in the same time.

[†] *Hint*: $y =$ Average velocity × time $= \frac{1}{2}(u+v)t$, and t is equal to the velocity change $(v-u)$ divided by the acceleration g.

32 *Fields*

(e) Suppose the initial height of the object above the ground was h_1 and the final height was h_2, so that $y = h_1 - h_2$. Show that

$$mgh_1 + \frac{1}{2}mu^2 = mgh_2 + \frac{1}{2}mv^2. \tag{2.1}$$

Our working has shown us that there is a conserved quantity in our motion. While the velocity and the height change, the total of $mg \times \text{height} + \frac{1}{2}m \times \text{speed}^2$ remains the same. We call this conserved quantity energy. The part of the energy which depends on the speed ($\frac{1}{2}mv^2$) is the energy of motion or *kinetic energy*.

The other part (mgh) is called the *potential energy* because although we cannot see it, an object at height could develop considerable motion if it were allowed to fall.* You may ask, 'Where is this energy stored?' Is it in the object? Or in the Earth which set up the gravitational field in the first place? Or is it shared, and if so how? Our most helpful answer is to say that the energy is stored in the gravitational field itself.

You will notice that if an object is lifted, its potential energy will change by an amount equal to its weight (mg) multiplied by the height by which it is lifted (h). When we allow this force to change the height of the object (i.e. by letting it go) we say that the force does work on the object, and we define the 'work done' as the product of the force and the distance the object moves in that direction.

Work done by force = Force × Distance moved by object in that direction.

Note: If the object is moving in the opposite direction to the force (as when a car is being slowed down by its brakes), the work done by the force is negative and the kinetic energy will be reduced.

The 'work done' turns out to be the same as the amount of energy transferred from one form to another. Here the work done by the force on the object (mgy) is equal to the amount of energy 'converted' from potential to kinetic in form.

Having solved our problem of analysing motion in one dimension, we need to extend this to other dimensions. However, there is a vital tool we shall need first, and this is covered in the next workshop.

2.3 Workshop: Scalar product of vectors

Unlike real or complex numbers (to be dealt with in Chapter 4), there is no unique way of multiplying two vectors together. While the magnitudes of vectors can be multiplied, each vector also has a direction associated with it, and it is by no means obvious how to multiply the directions together.

It transpires that there are two multiplication procedures which can be applied to vectors. One gives its result as a scalar and is called the *scalar product*. The other gives its result as a vector and is called the *vector product*. When we need to write these down, the scalar product is written with a dot: **a** · **b**, while the vector product is written with a cross: **a** × **b**. For this reason, the scalar product is sometimes called

---

*In common speech, the word *potential* always means 'could develop' (e.g. 'He is a player with potential' = 'He could develop into a good player') and the same is true here.

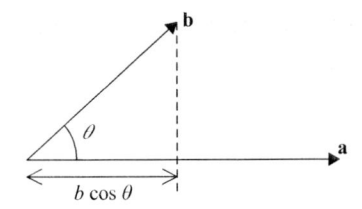

Fig. 2.3

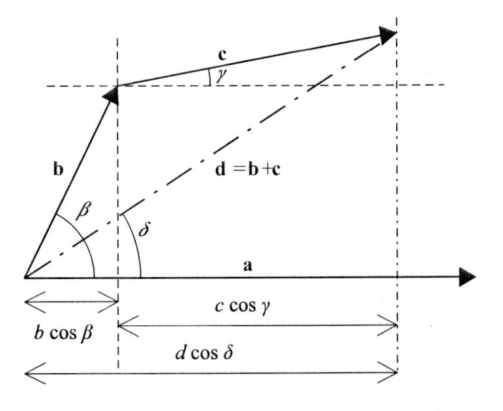

Fig. 2.4

the dot product, while the vector product is called the cross product. Vector products will be dealt with in a workshop in Section 3.1.3, and scalar products are introduced here.

The scalar product of two vectors $\mathbf{a} \cdot \mathbf{b}$ is defined as the magnitude of $\mathbf{a}$ multiplied by the component of $\mathbf{b}$ which is parallel to $\mathbf{a}$.

In Figure 2.3, the component of $\mathbf{b}$ parallel to $\mathbf{a}$ is equal to $b \cos \theta$. Accordingly:

$$\mathbf{a} \cdot \mathbf{b} = ab \cos \theta, \qquad (2.2)$$

where we use the italic letter a to represent the magnitude of vector $\mathbf{a}$. Notice that when you take the scalar product of a vector with itself, the angle $\theta = 0$, and so $\mathbf{a} \cdot \mathbf{a} = a^2$.

(a) Show that the following definition of $\mathbf{a} \cdot \mathbf{b}$ is equivalent to the one above: $\mathbf{a} \cdot \mathbf{b}$ is defined as the magnitude of $\mathbf{b}$ multiplied by the component of $\mathbf{a}$ which is parallel to $\mathbf{b}$. One consequence of your reasoning is that $\mathbf{a} \cdot \mathbf{b} = \mathbf{b} \cdot \mathbf{a}$.

(b) If $\mathbf{a} = \begin{pmatrix} a_x \\ 0 \end{pmatrix}$ and $\mathbf{b} = \begin{pmatrix} b_x \\ b_y \end{pmatrix}$, show that $\mathbf{a} \cdot \mathbf{b} = a_x b_x$.*

*Hint: Imagine the situation in Figure 2.3 again, but with vector $\mathbf{a}$ lined up along the x-axis.

34 *Fields*

(c) Evaluate $\mathbf{a} \cdot \mathbf{b}$ if $\mathbf{a} = \begin{pmatrix} 0 \\ a_y \end{pmatrix}$ and $\mathbf{b} = \begin{pmatrix} b_x \\ b_y \end{pmatrix}$.

(d) The aim of this part is to convince you that just as in normal algebra where $a(b+c) = ab + ac$, with scalar products $\mathbf{a} \cdot (\mathbf{b} + \mathbf{c}) = \mathbf{a} \cdot \mathbf{b} + \mathbf{a} \cdot \mathbf{c}$. Look at Figure 2.4, where $\mathbf{d}$ is defined as the vector sum of $\mathbf{b}$ and $\mathbf{c}$, and the angles made by $\mathbf{b}$, $\mathbf{c}$, and $\mathbf{d} = \mathbf{b} + \mathbf{c}$ with the direction of $\mathbf{a}$ are denoted β, γ, and δ.

   (i) Write $\mathbf{a} \cdot \mathbf{b}$ in terms of a, b, and an angle.
   (ii) Write $\mathbf{a} \cdot \mathbf{c}$ in terms of a, c, and an angle.
   (iii) Write $\mathbf{a} \cdot \mathbf{d}$ in terms of a, d, and an angle.
   (iv) Write $\mathbf{a} \cdot \mathbf{b} + \mathbf{a} \cdot \mathbf{c}$ as the sum of your answers to parts (i) and (ii).
   (v) Show, by reference to Figure 2.4 that your answer to (iv) is the same as your answer to (iii) and as such is equal to $\mathbf{a} \cdot (\mathbf{b} + \mathbf{c})$.

(e) Armed with your knowledge from part (d), and given that $\begin{pmatrix} a_x \\ a_y \end{pmatrix} = \begin{pmatrix} a_x \\ 0 \end{pmatrix} + \begin{pmatrix} 0 \\ a_y \end{pmatrix}$, combine your answers to parts (b) and (c) to show that $\begin{pmatrix} a_x \\ a_y \end{pmatrix} \cdot \begin{pmatrix} b_x \\ b_y \end{pmatrix} = a_x b_x + a_y b_y$.

It turns out that our reasoning in part (e) can be extended to as many dimensions as you like. In three dimensions, $\mathbf{a} \cdot \mathbf{b} = a_x b_x + a_y b_y + a_z b_z$.

2.4 Workshop: Motion in a uniform field in three dimensions

In some senses, our one dimensional work in Section 2.2 on falling objects needs no extension to three dimensions since the other two dimensions are irrelevant. Any non-vertical motion will not change the potential energy and will therefore not change the kinetic energy (or speed). Put in other words, non-vertical motion does not change the work done by the force on the object. Thus equation (2.1) remains true in three dimensions as long as we let the letters h continue to represent the heights of objects without reference to horizontal position. Interestingly, we can let the letters u and v (which used to represent the vertical components of velocity) now represent three dimensional speeds without putting a spanner in the works.*

That said, we do wish to be in a position to analyse three dimensional motion using vectors without particular regard for aligning our coordinates so that one points downwards, and so we extend our analysis.

We now return to the situation of Section 2.2, but extend our reasoning to all three dimensions. Imagine an object of mass m is subject to a uniform acceleration $\mathbf{g}$, and starts with velocity $\mathbf{u}$. At time t later the object has velocity $\mathbf{v}$ and has a displacement $\mathbf{s}$ with respect to its initial position. Thus the variables $\mathbf{s}$, $\mathbf{g}$, $\mathbf{u}$, and $\mathbf{v}$, are all vectors.

(a) The acceleration $\mathbf{g}$ is equal to the rate of change of velocity, so $\mathbf{g} = $ velocity change $\div$ time $= (\mathbf{v} - \mathbf{u})/t$. The displacement $\mathbf{s}$ will still be given by average

---

*How is this justified? The velocities in equation (2.1) are vertical velocities which we now call u_z and v_z. Is it OK to replace these v_z^2 and u_z^2 with the squares of the speeds v^2 and u^2? Yes. The length of a vector (x, y, z) is given by Pythagoras Theorem as $x^2 + y^2 + z^2$. Similarly the square of the speed of the object will be given by $v^2 = v_x^2 + v_y^2 + v_z^2$. Neither v_x nor v_y will change (there are no forces acting in these directions) and so $v^2 - u^2 = v_z^2 - u_z^2$.

velocity × time, and hence $\mathbf{s} = \frac{1}{2}(\mathbf{u}+\mathbf{v})t$. We can combine these two equations to eliminate t by calculating the scalar product $\mathbf{g}\cdot\mathbf{s}$. Calculate this product, and show that it is equal to $\frac{1}{2}(v^2 - u^2)$ where u^2 means the square of the speed and is equal to $\mathbf{u}\cdot\mathbf{u}$.

In Section 2.2, we defined the work done by a force on an object as the magnitude of the force multiplied by the distance the object moves in the same direction as the force. This is naturally written using vector notation as

$$\text{Work done by force} = \mathbf{F}\cdot\mathbf{s}. \qquad (2.3)$$

(b) Use equation (2.3) together with your answer to part (a) to show that the work done by the force of gravity $\mathbf{F} = m\mathbf{g}$ on the object is equal to the gain in the object's kinetic energy.

When work is done by gravity on the object, it *loses* potential energy. This means that the change in potential energy of the object is $-m\mathbf{g}\cdot\mathbf{s}$. This minus sign is vital. If you want to raise the potential energy of the object, you need to give it a displacement in the opposite direction to the field strength $\mathbf{g}$ (i.e. upwards).

Frequently we wish to make comments about how 'rich' a particular position is in terms of potential energy without specifying the mass of the object. To help us do this, we define the *potential* of a point as the potential energy per unit mass. The change in potential accompanied by a change in displacement $\mathbf{s}$ is given by

$$\Delta\phi = -\mathbf{g}\cdot\mathbf{s}, \qquad (2.4)$$

where we use the symbol ϕ for potential.

(c) The analysis so far in this chapter has concentrated on uniform gravitational fields. Electrostatic fields are just as straightforward, but we use $\mathbf{E}$ to represent the electric field strength. The force on a charge is then given by $\mathbf{F} = q\mathbf{E}$. Notice that $\mathbf{E}$ points in the direction in which a *positive* charge would experience a force.

Rework the analysis above, with the acceleration now being given by $\mathbf{F}/m = q\mathbf{E}/m$. Show that the change of potential energy of an object is now given by $-q\mathbf{E}\cdot\mathbf{s}$, and that the change in potential (i.e. potential energy per unit charge) when we move a charge by displacement $\mathbf{s}$ is now,

$$\Delta\phi = -\mathbf{E}\cdot\mathbf{s}. \qquad (2.5)$$

(d) A capacitor consists of two parallel metal plates separated by a distance d, one of which is charged positively, while the other is negatively charged. Assume that the electric field ($\mathbf{E}$) is uniform in the region between the plates and points towards the negative plate. Show that the change in potential when moving from a point on the negative plate to any point on the positive plate* is the

---

Hint: It may help if you assume that the charge is taken on a straight line route from the one point to the other.

36 *Fields*

same and is equal to Ed. Given that this is equal to the potential energy gained per unit charge, this will be the same as the voltage difference between the plates.

2.5 Non-uniform fields

All of the analysis up to now has concentrated on uniform fields, where **g** or **E** remains the same wherever you go. But what happens if they change? In practice they could change for one of two reasons. Either the field everywhere could change (i.e. **E** could be a function of time), or the field could be time-independent but varying from place to place. In either case, a moving particle would experience different strengths or directions of field along its route.

Our first observation is that if the displacement is sufficiently small (and the time taken to move correspondingly short), the field will not change appreciably over the course of this mini-motion. As in Chapter 1, we denote this mini-motion with the letter δ so here the small displacement is $\delta \mathbf{s}$. We say that the minuscule amount of work done by the force δW is

$$\delta W = \mathbf{F} \cdot \delta \mathbf{s}. \tag{2.6}$$

Correspondingly, the change in the potential will be

$$\delta \phi = -\mathbf{g} \cdot \delta \mathbf{s} \tag{2.7}$$

for a gravitational field, or

$$\delta \phi = -\mathbf{E} \cdot \delta \mathbf{s} \tag{2.8}$$

for an electrostatic one.

Notice that if we divide both sides of (2.6) by the small amount of time taken δt, we get an expression for the power P – that is, the rate at which the force is working on the object:

$$P = \frac{\delta W}{\delta t} = \frac{\mathbf{F} \cdot \delta \mathbf{s}}{\delta t} = \mathbf{F} \cdot \frac{\delta \mathbf{s}}{\delta t} = \mathbf{F} \cdot \mathbf{v}. \tag{2.9}$$

Thus power is the scalar product of force with velocity.

This is all very interesting, but of course we will want to analyse motions which are not minuscule. Our approach is to break our intended motion $\Delta \mathbf{s}$ into lots of very small pieces $\delta \mathbf{s}$. We then work out the work done by each of the small pieces using equation (2.6), and then add them up to get the total work done ΔW:

$$\Delta W = \sum \delta W = \sum \mathbf{F} \cdot \delta \mathbf{s}. \tag{2.10}$$

But how small should the pieces be? As small as possible, of course, and so we let our summation in equation (2.10) become an integral*:

$$\Delta W = \int \mathbf{F} \cdot d\mathbf{s}. \tag{2.11}$$

This may be your first example of what is called a *line integral*, since it is executed along the line of the path (all the little stages $d\mathbf{s}$). Before we explain in greater detail how to work this out, let us note in passing that equivalent logic with our potential functions gives

$$\Delta \phi = - \int \mathbf{g} \cdot d\mathbf{s} \tag{2.12}$$

for gravitational fields, and

$$\Delta \phi = - \int \mathbf{E} \cdot d\mathbf{s} \tag{2.13}$$

for electrostatic ones.

2.6 Workshop: Evaluating line integrals

In this workshop we shall outline a few methods of evaluating the integrals, starting with the simplest cases.

Sometimes you can choose the route of the integral so that the field strength E, and the angle it makes to the route θ are the same all along the path. In this case, the equation simplifies to

$$\int \mathbf{E} \cdot d\mathbf{s} = \int E \cos\theta \, ds = E \cos\theta \int ds = ES \cos\theta,$$

where S is the total length of the path.

(a) The magnetic field round a straight current-carrying wire points round the wire in a set of circles centred on, and perpendicular to, the wire. It has been demonstrated that the integral of the magnetic field strength round any loop enclosing the wire, and coming back to the same place $\oint \mathbf{B} \cdot d\mathbf{s} = \mu_0 I$, where I is the current in the wire.[†]

If we assume that all points equidistant from the wire have equally strong magnetic flux B, show by performing the line integral round a circle of radius r centred on the wire and perpendicular to it that $2\pi r B = \mu_0 I$, and thus find an equation for B.

Before leaving this method, it is worth pointing out that you are allowed to choose the route of the integral in certain circumstances only. We shall deal with this later in Section 2.7. If in doubt, do not change the route.

---

*See Section 1.1.2 for an introduction to what this means. Further details on how to use them are given in a special workshop in Section 7.1.

[†]The symbol $\oint$ with a ring around it simply means that the integral has its start and end points in the same place. Sometimes this kind of line integral is called a loop integral, since the path of integration takes the form of a loop.

38 Fields

Our next approach is to declare a new scalar variable s, equal to the distance travelled along the path from the start point of the integral to the end. By knowing the field $\mathbf{E}$ at each point along the route, we can rewrite $\mathbf{E}$ as a function of s. Once this is done, we can work out the angle $\theta(s)$ between the vector $\mathbf{E}(s)$ and the direction of the path for each point on the path. We can then write

$$\int \mathbf{E} \cdot d\mathbf{s} = \int \mathbf{E}(s) \cdot d\mathbf{s} = \int E(s) \cos \theta(s) \, ds,$$

which is an ordinary integration of a function of s with respect to s, and as such can be integrated using the methods you will already be familiar with.

(b) As you may know, the magnitude of the electric field in the vicinity of a point charge Q is given by $E = Q/4\pi\varepsilon_0 r^2$ where r is the distance of the point from the charge Q and ε_0 is a constant called the permittivity of free space. The direction of the field is outwards (away from Q) if the charge is positive and inwards if the charge is negative. Calculate the potential difference between a point R away from Q and a point twice as far away by integrating $\Delta\phi = -\int \mathbf{E} \cdot d\mathbf{s}$ along a radial line from $r = R$ to $r = 2R$.

Our final method is to write the three components of $\mathbf{E}$ separately. If E_x is the x-component of $\mathbf{E}$, then we can use part (e) of Section 2.3 to rewrite our integral as

$$\int \mathbf{E} \cdot d\mathbf{s} = \int E_x dx + E_y dy + E_z dz = \int E_x dx + \int E_y dy + \int E_z dz.$$

The integral has now become a sum of three 'ordinary' integrals. If we can arrange for one of the coordinate axes (say, x) to be parallel to the route of integration, then two of the integrals sum to zero since there is no range of y or z to be integrated over. Just watch out: E_x is the x-component of $\mathbf{E}$, and will in general still be a function of x, y, and z (not just x).*

(c) Suppose our field $\mathbf{E}$ is given by

$$\mathbf{E} = \frac{F}{a^2} \begin{pmatrix} yz \\ xz \\ xy \end{pmatrix},$$

where F and a are constants. Calculate the potential difference between the origin and the point $(1,1,1)$ along the straight line given by the formula (s, s, s) where the variable s runs from 0 to 1.

(d) Repeat part (c) but now perform the line integral in three stages – from $(0,0,0)$ to $(1,0,0)$, then on to $(1,1,0)$, and then from there to $(1,1,1)$. For this particular choice of field, you should get the same answer.

---

*When evaluating $\int E_x(x,y,z)dx$, write y and z as functions of x using your knowledge of the shape of the path or route you are integrating over. After all, if you know the value of x of a point on a known path, you should be able to work out the values of y and z as well. [If you cannot, then you set up your coordinates badly, and should try again.]

(e) In Section 2.2, we proved that the total of the potential and the kinetic energy was constant for a particle moving in a uniform field in one dimension. In Section 2.4, we generalized this to three dimensions. Here we will present a proof of the same result applicable to three dimensional non-uniform fields.

  (i) Using the product rule for differentiation and information from our workshop on scalar products (Section 2.3), prove that
  $$\frac{d}{dt}v^2 = 2\mathbf{v} \cdot \frac{d\mathbf{v}}{dt} = 2\mathbf{v} \cdot \mathbf{g},$$
  where the acceleration $d\mathbf{v}/dt$ is equal to $\mathbf{g}$ because the gravitational force is the only force acting on the object.

  (ii) Our work in Section 2.4 taught us that the change in potential energy of the object is given by
  $$-\int m\mathbf{g} \cdot d\mathbf{s} = -\int m\mathbf{g} \cdot \frac{d\mathbf{s}}{dt} dt = -\int m\mathbf{g} \cdot \mathbf{v}\, dt,$$
  where we have taken the liberty of using a substitution to change the variable of integration to the scalar t. Use your answer to (i) to show that this is the same as
  $$-\int \frac{m}{2} \frac{d(v^2)}{dt} dt = -\left[\frac{1}{2}mv^2\right],$$
  and hence the gain in potential energy is still equal to the loss of kinetic energy.

(f) In all of our work so far, we have assumed that only the gravitational field caused a force to act on the particle, and as a result the acceleration of the particle $\mathbf{a}$ was equal to $\mathbf{g}$. This may have given you some concern that our reasoning was limited to this case. To prove that it is not, consider a particle moving in a gravitational field $\mathbf{g}$ which is also being pulled around by an additional force $\mathbf{F}$. We shall show that the work done by the force $\mathbf{F}$ on the system is equal to the sum of the gains of the potential and kinetic energy.

  (i) By considering the resultant force on the particle, show that $\mathbf{F} = m(\mathbf{a} - \mathbf{g})$.
  (ii) The work done by the additional force is $\int \mathbf{F} \cdot d\mathbf{s}$. Use your result to (i) to show that this is equal to $m \int \mathbf{a} \cdot d\mathbf{s}$ + the gain in potential energy.
  (iii) Using the methodology of part (e), show that $m \int \mathbf{a} \cdot d\mathbf{s}$ is the gain in kinetic energy of the object.

Accordingly we have shown that the energy inserted into the system by the action of force $\mathbf{F}$ has been shared between the potential and kinetic energy of the system, and that the total energy has been conserved.

2.7 Potential gradients

We have spent the majority of the sections in this chapter defining potentials and giving methods for their calculation. Apart from a knowledge of the energy of the system, what was the point? One answer is that the potential is a scalar and is much

40 Fields

[Figure with labels: "Gradient = ∂h/∂y", "Gradient = ∂h/∂x", "$\frac{\partial h}{\partial y}\delta y$", "$\frac{\partial h}{\partial x}\delta x$", "Total height gain", "δy", "δx"]

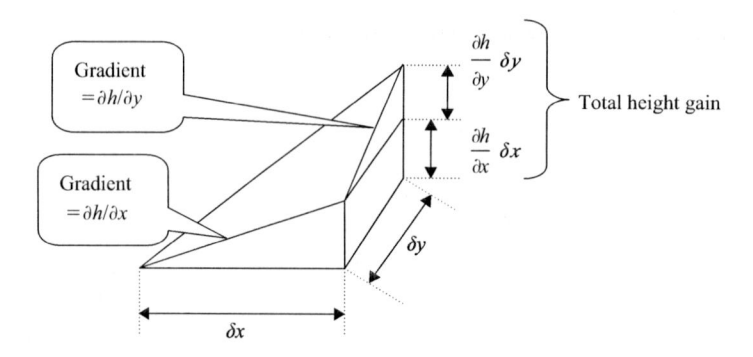

Fig. 2.5

easier to work with than three dimensional vectors. Accordingly if we succeed in writing all of the information about a field into a potential function ϕ, we simplify many of the calculations we might wish to do with it.

To do so, we first limit ourselves to the fields whose potentials ϕ are functions of position alone. This means that the potential difference between two points has a single value, and so when the potential difference is calculated by performing the integral $\Delta\phi = -\int \mathbf{g} \cdot d\mathbf{s}$, you get the same answer no matter what route you use to get from the one point to the other.

In Section 2.2, we have already ascertained that the potential energy of an object in a uniform, downwards gravitational field is given by mgh where h is the height of the point, and so the potential will be given by $\phi = gh$.

Imagine that you are in hilly terrain, and that instead of a map with contours, you are given a mathematical function to tell you the height of the ground at any point. If we set up coordinate axes where x points east, y points north, and z points upwards, this means that we have a function $z = h(x, y)$.

If I ask the question, 'How steep is the ground here?' the answer depends on the direction I am walking. The gradient of the slope as measured by someone walking east (the $+x$-direction) is written $\partial h/\partial x$, which means the derivative of h with respect to x while keeping y constant. The gradient of the slope as measured by someone walking north is given by $\partial h/\partial y$.*

If I walk in some other direction, when I move a distance $\delta\mathbf{s} = (\delta x, \delta y)$, it is as if I moved δx east and then moved δy north. This situation is shown in figure 2.5. Accordingly the change in the ground's height as I walk is

$$\delta h = \frac{\partial h}{\partial y}\delta y + \frac{\partial h}{\partial x}\delta x,^\dagger \tag{2.14}$$

---

*These derivatives written with curly ∂ are called partial derivatives, and they crop up whenever you have a function with more than one independent variable.

†We are assuming that our displacement δs is sufficiently small that the gradients do not change appreciably as we walk this tiny distance.

and so the change in my gravitational potential is

$$\delta\phi = \delta(gh) = g\delta h = \frac{\partial \phi}{\partial y}\delta y + \frac{\partial \phi}{\partial x}\delta x. \tag{2.15}$$

If we now write the displacement I have moved as a vector $\delta\mathbf{s} = (\delta x, \delta y)$, I can write my change in gravitational potential as $\delta\phi = \mathbf{D} \cdot \delta\mathbf{s}$, if the vector $\mathbf{D} = (\partial\phi/\partial x, \partial\phi/\partial y)$. This vector gives me all the information I need to work out the gradient when walking in any direction on the hillside.

If I choose to walk in the direction of $\mathbf{D}$, then I choose the steepest uphill route (I get the biggest change in ϕ for a particular value of δs). If I walk in the opposite direction to $\mathbf{D}$, I choose the steepest downhill route. If I walk perpendicularly to $\mathbf{D}$, then $\mathbf{D} \cdot \delta\mathbf{s} = 0$, and I neither rise nor fall. My potential energy does not change – I am walking along the contour line.

We now extend our logic to motion in any kind of gravitational field where the potential ϕ is given by a known function of x, y, and z. The change in potential after a mini-motion $\delta\mathbf{s}$ is given by equation (2.7) as

$$\delta\phi = -\mathbf{g} \cdot \delta\mathbf{s} = -(g_x\,\delta x + g_y\,\delta y + g_z\,\delta z). \tag{2.16}$$

The change in potential is also given by equation (2.15), where we have adapted the equation to take into account motion in three dimensions:

$$\delta\phi = \frac{\partial \phi}{\partial x}\delta x + \frac{\partial \phi}{\partial y}\delta y + \frac{\partial \phi}{\partial z}\delta z. \tag{2.17}$$

Comparing equations (2.16) and (2.17), it follows that

$$g_x = -\frac{\partial \phi}{\partial x}, \quad g_y = -\frac{\partial \phi}{\partial y}, \quad g_z = -\frac{\partial \phi}{\partial z}. \tag{2.18}$$

This is more conveniently written as $\mathbf{g} = -\nabla(\phi)$, where ∇ is a vector function, which acts on the scalar potential ϕ to produce a vector with the partial derivatives of ϕ in each of the three directions as its components. Given that its three components are the gradients of ϕ in the directions of the three axes, it is called the gradient function (or grad for short).

We now summarize our two most important relationships linking field $\mathbf{g}$ and potential ϕ:

For calculating $\Delta\phi$,

$$\Delta\phi = -\int \mathbf{g} \cdot d\mathbf{s}. \tag{2.19}$$

42 Fields

For calculating **g**,

$$\mathbf{g} = -\nabla(\phi) = \begin{pmatrix} -\dfrac{\partial \phi}{\partial x} \\ -\dfrac{\partial \phi}{\partial y} \\ -\dfrac{\partial \phi}{\partial z} \end{pmatrix}. \tag{2.20}$$

Thus we use line integrals to work out potentials from a knowledge of the field, and we use the gradient function to work out the field from a knowledge of the potential.

Not all fields can be described using a potential in this way. For some functions **g**, there are no functions ϕ which can be found and which will satisfy equation (2.20). However, gravitational and electrostatic* fields *can* be described using potentials as shown above. These gravitational or electrostatic potential functions are functions of position only, and accordingly it does not matter which route you take between two fixed points: the total change in potential will be the same. Such fields are said to be conservative fields, and their field lines have no loops in them (see Q9 for more details).

Q4 A particle is held to a point on a horizontal surface using a spring. The force on the particle has magnitude $F = k\sqrt{x^2 + y^2}$, and is directed towards the point $x = 0$, $y = 0$.
   (a) What is the potential energy of the system as a function of x when the particle moves on the line $y = 0$? Take the potential energy to be zero when the particle is at $x = 0$.
   (b) What is the potential energy of the system as a function of x and y when the particle can be anywhere on the plane? +
   (c) Write the force acting on the particle at point (x, y) as a vector.
   (d) Calculate the partial derivatives of your answer to (b) with respect to x and y, and thus verify your answer to (c).

Q5 If the potential function for a field is $\phi = 3x^2 + 2yz + 9xy/z$, work out the force **F** acting on a particle in this field as a function of its position.

Q6 For the following forces, attempt to find a potential function to describe their fields
   (a) $\mathbf{F} = (0,\ 65,\ 2)$
   (b) $\mathbf{F} = (54y,\ 54x,\ 3)$
   (c) $\mathbf{F} = (54y,\ 59x,\ 3z)$
   (d) $\mathbf{F} = (54y,\ -54x,\ 2x)$
   (e) $\mathbf{F} = (-3x,\ -3y,\ -3)$.

Q7 The height of a point on a hillside is given by $h = 0.06x - 0.02y + 100$ m, where x and y are the east and north coordinates of the point.
   (a) Calculate $\partial h/\partial x$. What does this mean?

---

*By electrostatic field, we mean the sort of electric field set up by charged objects. Electric fields can also be produced by chemical or electromagnetic means; however, our discussion here is limited to the electrostatic case, ensuring that we deal with a conservative field.

(b) Calculate $\partial h/\partial y$. What does this mean?
(c) How can you tell that this surface is not curved?
(d) In which direction is the gradient steepest?
(e) What is the steepest gradient of the slope?
(f) Which way would I walk to keep at the same height?
(g) I walk from (100,200) to (300,800). Calculate the change in height of my position using the equation $h = 0.06x - 0.02y + 100$ m, and also using your answers to (a) and (b). Check that your answers agree.

Q8 We shall show in Section 2.8.3 that the gravitational field due to a point mass M at the origin is given by
$$\mathbf{g} = -\frac{GM}{r^2}\hat{\mathbf{r}},$$
where $\hat{\mathbf{r}}$ is a unit vector pointing away from the origin.

(a) Satisfy yourself that in Cartesian coordinates:
$$\hat{\mathbf{r}} = \frac{1}{\sqrt{x^2 + y^2 + z^2}} \begin{pmatrix} x \\ y \\ z \end{pmatrix}.$$

(b) Show that in Cartesian coordinates, $\mathbf{g}$ can be written:
$$\mathbf{g} = -\frac{GM}{(x^2 + y^2 + z^2)^{3/2}} \begin{pmatrix} x \\ y \\ z \end{pmatrix}.$$

(c) Find the potential function which gives this field. That is, find ϕ such that $-\nabla\phi = \mathbf{g}$. If you are stuck, look it up in the hint below, and then verify that it is correct.*

Q9 A circular loop of wire is placed in a changing magnetic field. The electric field set up by the moving magnet is given by $\mathbf{E} = (-Ay, Ax, 0)$, where A is a constant.

(a) Show that it is impossible to write $\mathbf{E}$ as the gradient of a scalar function ϕ.
(b) Calculate the value of the loop integral $\oint \mathbf{E} \cdot d\mathbf{s}$ where S is a circular path in the $x - y$ plane, centred on the origin, with radius r, and show that it is non-zero. Your answer gives the reading a voltmeter would show if connected to a single turn of wire: it is the voltage induced in the wire. +

Your answer to (b) shows that an electron moving round the wire and coming back to its starting place comes back with a different amount of potential energy. How can this be possible? The answer is that it has picked up or lost energy by interacting with the changing magnetic field on the way – this is how generators work. We call this kind of field non-conservative, since the potential of a point is not a unique function of its position, but depends on the route of evaluation of the integral. In general all fields which have a tendency to produce loops in their field lines are non-conservative, and

*The function you are looking for is $\phi = -\dfrac{GM}{r} = -\dfrac{GM}{\sqrt{x^2 + y^2 + z^2}}$. By taking partial derivatives of this with respect to x, y, and z, you should be able to recover the three components of the $\mathbf{g}$ field given in the question.

44 *Fields*

they include magnetic fields and the electric fields made when magnetic fields change. Conservative fields (e.g. gravitational and electrostatic fields) can be distinguished in two ways – the field strength can be written as $-\nabla\phi$ and the field lines never form loops.

2.8 Setting up a field

At the beginning of the chapter, we commented that forces caused accelerations, and pondered what caused the forces. We have now developed the idea that fields cause forces. But what causes the fields? The answer is that masses set up gravitational fields, and charged particles set up electrostatic fields.

To quantify these interactions, we need to have an understanding of how these charges and masses set up fields. Once this is done, we have a complete picture of how an interaction can be modelled. In this section, we shall concentrate on electrostatic fields; since these are easier to manipulate in the laboratory, however, nearly all of the following reasoning also applies to gravitational fields.

One helpful way of visualizing an electric field is with field lines. No doubt you have done things in the past like plotting the magnetic field of a bar magnet, and we use a similar approach to drawing the field lines of an electric charge. The field lines are not real (but then again, the field itself is only a model), but by visualizing their flow we can picture what is going on much better as shown in figure 2.6.

As with the magnetic fields, the field lines give the direction of the field (the direction of **E**) at any point along their path. Given that the field cannot have more than one direction at any point, field lines never cross or branch out. Electrostatic field lines do not start or end in midspace. However, neither do they form complete loops like the magnetic field lines around a current-carrying wire or coil. The direction of the electric field is the direction a positive charge would experience a force. Accordingly, they must point away from positively charged objects and point towards negatively charged ones. Electric field lines therefore start on positive charges and always end on negative charges. Gravitational field lines always end on masses (Figure 2.7)

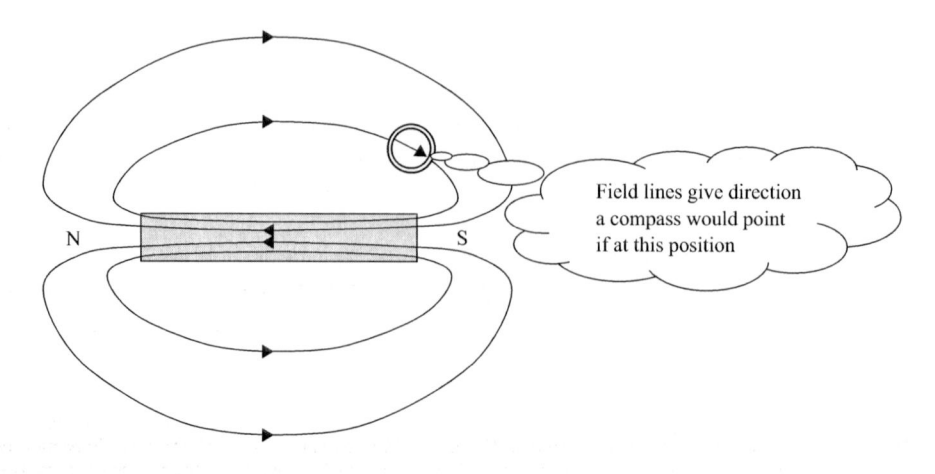

Fig. 2.6

Fig. 2.7

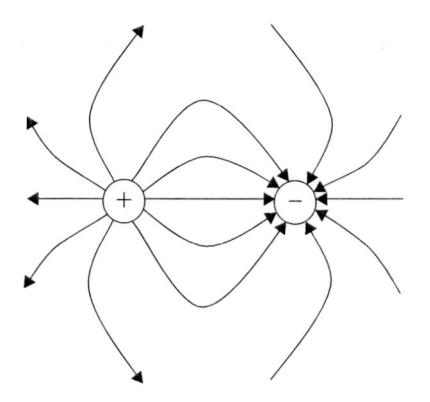

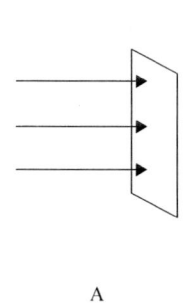

The electric field points in the direction a (+) charge would experience a force.

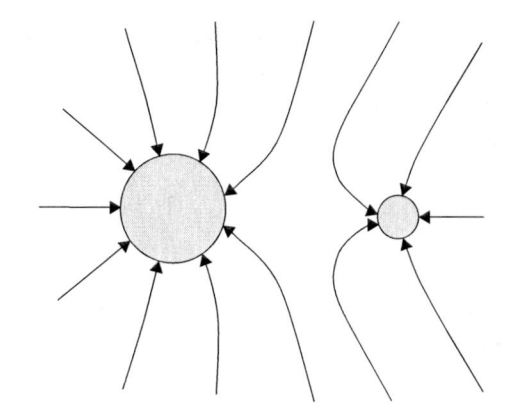

The gravitational field points in the direction a mass would experience a force.

Fig. 2.8

A Strength of field given by number of lines per unit area.

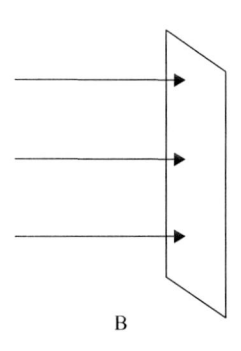

B This is a weaker field than A, since although it has the same 'number of lines', they are more spread out.

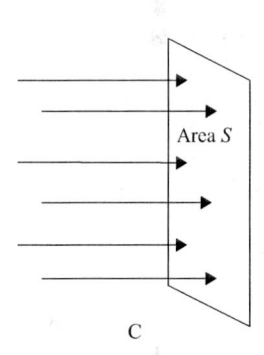

C 'Number of field lines' n defined so that $n/S = E$, the field strength.

Regions of stronger field stand out as areas where the field lines are closer together. In fact, we can build up a model where the electric field strength is represented by the 'density' of the field lines – that is the number of lines per unit area (where we count in a plane perpendicular to the field lines) as shown in figure 2.8.*

$$E = \frac{\text{Number of field lines}}{\text{Perpendicular area passed through}}. \tag{2.21}$$

A caution must be given at this point. Field lines are not discrete, and there is no 'real' way of counting them. However, if each field line represents a certain defined

---

*This gives us our working definition for the 'number of field lines' passing through a surface. The number need not be an integer, since there is nothing specific about each individual line. They merely give us a picture of what the field is doing. There is an analogy here with contour lines on a map – the map maker can put as many or as few as they see fit – a line every 5 m of altitude, a line every 100 m or every 1 mm – the one thing that would be useless for a map user is the truth of a infinite continuum of lines all smudged together.

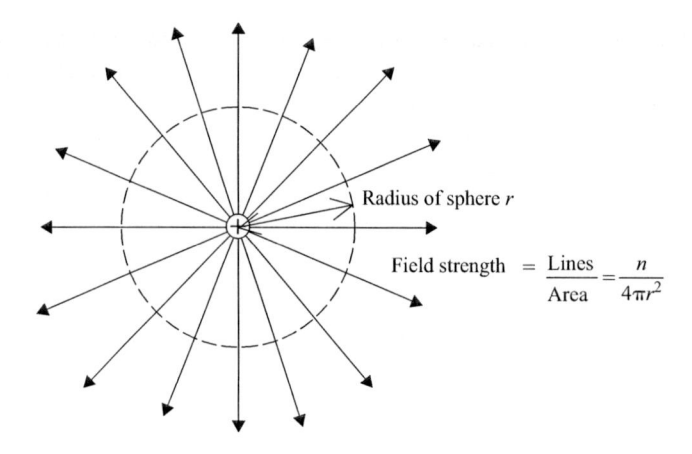

Fig. 2.9

amount of field, then counting them has some meaning in evaluating the total amount of field emanating from a charge or passing through an area. We shall define things more rigorously in terms of surface integrals in Section 2.8.4, however, until then we shall continue to use the visually helpful idea of 'number of field lines' with the proviso that it is just a convenient model for the complex business of field strength.

We can use this model to help us understand where Coulomb's law for the force between two charges comes from.

The field of a point charge $+Q$ (or any spherically symmetrical arrangement of charge) must, by symmetry, point inwards or outwards. If the charge is positive, the field lines stream outwards in the direction in which small positive charges would be repelled if placed there. Given that the field lines do not end, the lines carry on going to infinity. However, they spread out as they do so, and this indicates that the field gets weaker as we get further away from the positive charge in the centre as shown in figure 2.9.

If we draw a spherical shell centred on the charge $+Q$ with radius r, it will have area $4\pi r^2$. If the electric field strength E is given by the density of field lines, then it follows that the total number of field lines must be equal to the field strength multiplied by the perpendicular area through which they pass. Now our field is radial, and so will be perpendicular to the sphere. The total number of field lines is then given by

$$n = E \times 4\pi r^2.$$

This number must be the same no matter what the size of the sphere (assuming of course that the sphere is larger than the charged object itself). Therefore we can write:

$$E = \frac{n}{4\pi r^2},$$

and so the force experienced by a charge $+q$ at this distance will be

$$F = qE = \frac{nq}{4\pi r^2}.$$

If our charged object Q suddenly doubled in the amount of charge it stored, we would expect the field E and hence the force F to double too. It therefore seems sensible to say that the number of field lines streaming out from a positive charge is proportional to the charge on that object. We define the constant of proportionality to be the permittivity of free space ε_0, so

$$Q = n\varepsilon_0. \tag{2.22}$$

Our equations for the field strength and force then become

$$\begin{aligned} E &= \frac{Q}{4\pi\varepsilon_0 r^2} \\ F &= qE = \frac{qQ}{4\pi\varepsilon_0 r^2}, \end{aligned} \tag{2.23}$$

and the second equation is known as Coulomb's law for the force experienced between two point charges separated by distance r. Our equations so far have only given the strength or magnitude of the electric field, however, directionality is easy to add,

$$\begin{aligned} \mathbf{E} &= \frac{Q}{4\pi\varepsilon_0 r^2}\hat{\mathbf{r}}, \\ \mathbf{F} &= q\mathbf{E} = \frac{qQ}{4\pi\varepsilon_0 r^2}\hat{\mathbf{r}}, \end{aligned} \tag{2.24}$$

where $\hat{\mathbf{r}}$ is a unit vector pointing radially outwards from the charge Q. The electrostatic field then points parallel to $\hat{\mathbf{r}}$(outwards) if Q is positive and inwards if Q is negative, just as it should. Similarly $\mathbf{F}$, which gives the force experienced by the charge q because of the presence of Q, will point in the $\hat{\mathbf{r}}$ direction if Q and q have the same sign, but will point the other way if Q and q have opposite signs. Thus like charges attract, and opposite charges repel.

Q10 Calculate the electric field strength at a distance of 3 m from a 6 µC charge.

2.8.1 Workshop: The electrostatic field surrounding a charged wire

In this workshop we imagine a long straight wire carrying a static charge of λ per metre, and we attempt to work out the electric field experienced at a distance r from the wire.

Assuming that the wire is positively charged, the field lines are sketched in Figure 2.10 as they pass through a cylinder of length l and radius r centred on the wire. Notice that they all go out through the curved surface, not the flat circular ends. Also, because of the symmetry, the strength of the electric field at all points on the curved surface of the cylinder must be equal.

(a) Using our definition that the 'number of field lines' is equal to the electric field multiplied by the perpendicular area it goes through, show that the magnitude of the electric field is given by

$$E = \frac{n}{2\pi r l}.$$

48 Fields

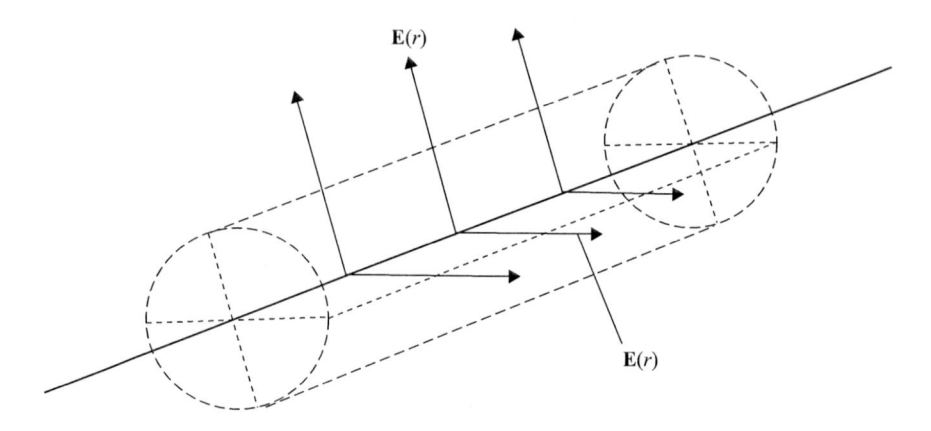

Fig. 2.10

(b) Using equation (2.22), derive another expression for n in terms of λ and ε_0.
(c) Equate your answers to (b) and (c) to give an expression for E which is not in terms of our fictitious field lines. You should get

$$E = \frac{\lambda}{2\pi\varepsilon_0 r}. \tag{2.25}$$

(d) Calculate the electric field strength at a distance of 2 cm from a long thin wire carrying a charge of 400 nC per metre of its length. What is the field strength 3 cm from the wire?
(e) A copper wire is cylindrical, with a cross-sectional area of 1 mm². Copper has a density of 8930 kg/m³, and each atom has a mass of 1.07×10^{-25} kg. Each atom has one electron free to move. Calculate the total charge on each metre of wire if all of these electrons were removed, and the electric field strength 200 m from the wire. Given that these electrons do all move when the wire carries a current, why are electric fields this big not formed during the passage of current?

2.8.2 Electrostatic charge in a parallel plate capacitor

Figure 2.11 shows a simple capacitor with plates of cross-sectional area A, and separation d. All of the electric field is contained within the two plates, and we shall assume that it is uniform. The 'number of field lines' n is clearly equal to EA using our definition (2.21).

If we apply our $Q = n\varepsilon_o$ rule to the positive plate, we see that $n = Q/\varepsilon_o$ and as $n = EA$, it follows that

$$E = \frac{Q}{\varepsilon_0 A}. \tag{2.26}$$

Setting up a field 2.8 49

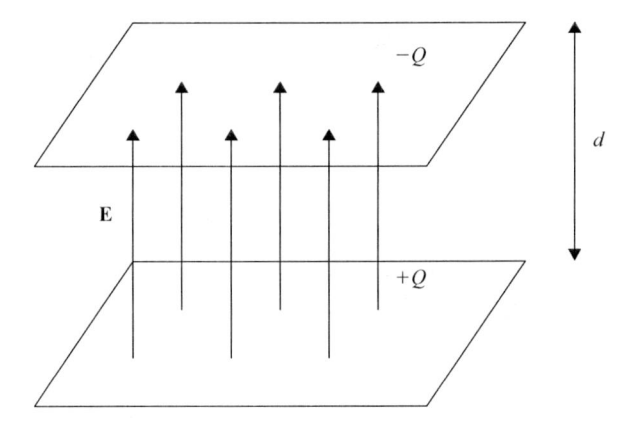

Fig. 2.11

In workshop 2.4 part (d), you showed that the voltage across the capacitor $V = Ed$. Given that the capacitance of a capacitor is defined as Q/V, the capacitance of the capacitor is equal to

$$C = \frac{Q}{V} = \frac{\varepsilon_0 E A}{Ed} = \frac{\varepsilon_0 A}{d}. \qquad (2.27)$$

Notice that if we put two such capacitors side by side so that they are effectively in parallel, then the area is doubled, and so is the capacitance. On the other hand, if two identical capacitors are put one above the other (in series) then the effective separation of the plates has doubled, so the capacitance is halved.

Q11 If I wanted to make a 1 F capacitor for the backup power supply in a computer using parallel plates, and I used plates with an area of 1 m^2, how close together must the plates be? In practice the plates are very thin, and are rolled into a cylinder to save space.

We can also use the capacitor to help us work out the energy content of an electric field. The energy stored in a capacitor is equal to $\frac{1}{2}CV^2$.* For our parallel plate capacitor, this equates to

$$\text{Energy} = \frac{1}{2}CV^2 = \frac{\varepsilon_0 A}{2d} \times (Ed)^2 = \frac{1}{2}\varepsilon_0 E^2 \times Ad. \qquad (2.28)$$

Now the volume of the capacitor is Ad, and thus the energy per unit volume is equal to $\frac{1}{2}\varepsilon_0 E^2$. While we have only proved this result for the uniform field in a parallel plate capacitor, any electric field can be thought of as a mosaic of tiny parallel plate

---

*The energy used to charge a capacitor to voltage V_f from 0 is given by $\int VI\,dt = \int V\frac{dQ}{dt}dt = \int V\,dQ = \int V\,d(CV) = C\int_0^{V_f} V\,dV = \frac{1}{2}CV_f^2.$

50 *Fields*

capacitors, and thus this formula actually applies to the energy density at any point in any electrostatic field.

2.8.3 Gravitational fields inside planets

We can apply the methods we have been developing to gravitational fields as well. When working with gravitational fields, we shall define our 'number of field lines' n analogously to (2.21) as the product of the field strength g with the perpendicular area through which it passes. Field lines end on masses, and we would expect the number of field lines ending on a mass to be proportional to the mass M itself. So that we end up with agreement with Newton's law of gravitation, we shall define a slightly complicated constant of proportionality:

$$n = 4\pi MG. \qquad (2.29)$$

We now apply our two ideas to a spherically symmetric distribution of mass. This ensures that the field will be radial at all points. If we draw a spherical surface within this distribution of radius r, and denote the total mass contained within this surface M, then $n = 4\pi MG$, and by multiplying the field strength g by the area of the sphere $4\pi r^2$, we find that $n = 4\pi g r^2$. Eliminating n between the two equations gives

$$g = \frac{GM}{r^2}, \qquad (2.30)$$

as we would have been expecting, since you will already know that the gravitational field of a point mass is given by this expression. However, there is more to this equation than meets the eye. It tells us that

- the mass outside our spherical surface has no contribution to the field within it.
- all of the mass within the spherical surface behaves gravitationally as if it were all concentrated at a single point at the very centre.

We inadvertently use this second point whenever we solve a problem involving orbits where we simply take the Earth to be a dot of mass 6.0×10^{24} kg at the centre of the orbit rather than the large planet we know it to be. While this approximation is not quite correct, the errors only come from the fact that the Earth is not spherically symmetric.

Q12 Assume the Earth to have a uniform internal density. At the planet's surface (6.4×10^6 m from the centre), the gravitational field strength is 9.8 N/kg. What do you expect the field strength to be at a point half way out (3.2×10^6 m from the centre)?

Q13 Prove that if a straight tunnel were constructed through the centre of the Earth, and if it were kept empty of magma and molten iron, and so on, that an object dropped into it would undergo simple harmonic motion about the centre of the Earth, and calculate the time period of the motion.*

Hint: You need to show that the force on the object is proportional to the distance from the centre of the Earth. If unfamiliar with harmonic motion, you may need to study Section 4.1.

Area measured perpendicular to field
= $\delta \mathbf{S} \cos \alpha$

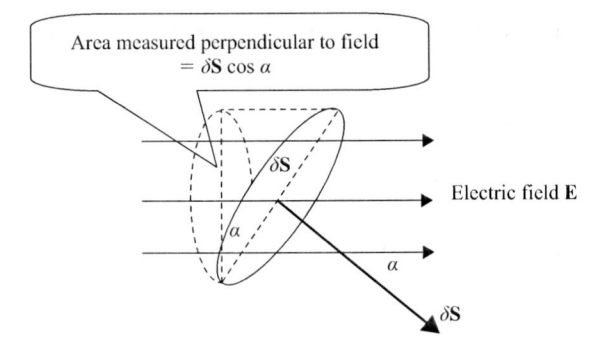

Electric field **E**

Fig. 2.12

Q14 Show that you still get simple harmonic motion with the same time period even if the tunnel does not go through the centre of the Earth, but rather joins any two points on the Earth's surface by a straight line.* Assume that the objects in the tunnel are able to pass up and down frictionlessly. +

2.8.4 Formalizing the notation

As mentioned at the beginning of this section, field lines are a fictitious pictorial representation of a field. Therefore we cannot really justify continuing further without formalizing what we mean by 'number of field lines'. After all, how do you count fictitious objects?

We need a more mathematical definition of what 'counting field lines' means. Suppose we take any closed surface which encloses the charges concerned, and we call this S. We can imagine this to be made up of lots of small pieces of surface each with area δS, and each of which is small enough to be almost flat. We assign a vector to each one called $\delta \mathbf{S}$, where the magnitude of the vector is equal to the area δS, and it points perpendicular to the surface (outwards) as shown in figure 2.12.

If $\delta \mathbf{S}$ makes an angle α to the field lines $\mathbf{E}$, then the area as measured perpendicular to the field is $\delta S \cos \alpha$. So the number of field lines crossing the area is given by $E \cos \alpha \, \delta S = \mathbf{E} \cdot \delta \mathbf{S}$. We may count the total number of lines n crossing the whole area S by adding up the contributions from each part. The result $n = \sum \mathbf{E} \cdot d\mathbf{S}$ is written as an integral as we break the area up into smaller and smaller pieces to fit the surface more accurately:

$$n = \oiint_S \mathbf{E} \cdot d\mathbf{S}, \qquad (2.31)$$

Hint: Here you need to show that the component of the force along the direction of the tunnel is proportional to the distance of the object from the tunnel's centre.

where the double integral reminds us that we are integrating over a two-dimensional area and the loop round the integrals reminds us that our surface is closed (it has no gaps or holes in it).

Earlier in equation (2.22), we defined the permittivity of free space ε_0 as the ratio of charge to the number of field lines produced Q/n. Using our definition of n, this becomes

$$\oiint_S \mathbf{E} \cdot d\mathbf{S} = \frac{Q}{\varepsilon_0}. \tag{2.32}$$

This equation is formally known as Gauss' law of electrostatics.

Any closed surface S without charge inside it must have a zero value of $\oiint_S \mathbf{E} \cdot d\mathbf{S}$ – this means that there must be as many field lines leaving the region as entering it. It follows that any charge-free area of space cannot have any field lines ending in it or starting from it.

For gravitational fields, we may use a similar definition of field lines and say that $n = \oiint_S \mathbf{g} \cdot d\mathbf{S}$. However, once we take directionality into account, we have to say that the number of field lines *leaving* an area is given by $n = -4\pi MG$ with the negative sign reminding us that lines come *into* the area in order to *end* on the mass. This gives us

$$\oiint_S \mathbf{g} \cdot d\mathbf{S} = -4\pi MG, \tag{2.33}$$

where M is the total mass contained within the surface S. It is worth mentioning that (2.33) is valid for all distributions of mass, not just spherically symmetric ones. However, it is only for the spherically symmetric distributions that we can simplify this expression to an inverse square law as in (2.30).

For magnetic fields, it is found that $\oiint_S \mathbf{B} \cdot d\mathbf{S} = 0$. Put into words, this means that magnetic field lines have no end. They may go from north to south outside of a bar magnet, but inside the magnet they all make their way back to north again.

Q15 Explain how the use of pole pieces (shaded grey in Figure 2.13) makes the magnetic field stronger in the gap in a horseshoe magnet.

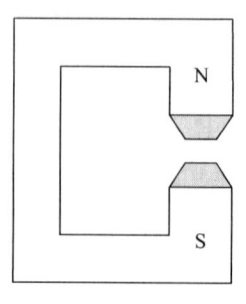

Fig. 2.13

Q16 Use Gauss' law of electrostatics to prove that in the absence of other charges, there is no electric field inside a uniformly charged hollow spherical shell. If we made a hollow region at the very centre of the Earth, what would be the gravitational field there?

Q17 Use Gauss' law of electrostatics to work out the magnitude of the electric field inside a uniformly charged sphere of radius R and total charge Q, if we make our measurement at a point a distance r away from the centre of the sphere. Find a potential function which will give this electrostatic field. +

2.9 Conclusion

After introducing the idea of a field as a description of the way space is affected by charges and masses and defining field strengths, we analysed motion in terms of distance rather than time. This led naturally to the ideas of work done and energy conservation. Through this we were able to develop a line integral for calculating the gain in potential energy of an object travelling on a route. For electrostatic, gravitational, and other conservative fields, the gain is independent of the route taken, and accordingly the potential of a point is a unique function of its position. The grad function can be used to calculate the field strength from a knowledge of the potential function.

We then moved on to discuss the way in which fields are set up, using the fictitious concept of counting field lines. This illustrated the similarities of electrostatic and gravitational fields and gave us a range of methods for calculating field strengths.

3
Rotation

In this chapter, we will examine the fundamental principles that are essential for a good understanding of rotational motion. We begin by considering the kinematics of rotating systems and the consequences of referring the physics of a situation to a rotating point of view (frame of reference). We then move on to ideas of the forces involved, which make up the 'dynamics' of this chapter. The techniques and ideas developed in the early part of this chapter are then applied to *the Kepler problem* in two workshops, which essentially outline the derivation of planetary trajectories through Newtonian mechanics. All the principles introduced in this chapter are also essential for a good understanding of the material contained in workshops in Sections 7.8–7.11 in Chapter 7. In order to fully appreciate those workshops, students seeing the material for the first time should work through Chapter 3 first.

3.1 Rotational kinematics and dynamics

3.1.1 Kinematics on a circular path

In this section (and accompanying workshops), we will only be looking at the kinematics of uniform circular motion, but many of the basics here form the foundation of more advanced work. We will be deriving a number of results that are essential to other chapters of this book, so it is worth making a mental note of this section and its results.

Figure 3.1 shows a rotating vector, or *radius vector*, $\mathbf{r}(t)$. The length of the vector, say r, does not change, but its direction is changing continuously. We can best represent this change in direction over time by defining a quantity called the *angular frequency*, ω. The angular frequency is most conveniently defined in terms of the angle in *radians* swept per unit time by the radius vector as it rotates. We will see why this is in a moment. For now, we define ω as follows:

$$\omega = \frac{2\pi}{T} = 2\pi f, \tag{3.1}$$

where T and f are the time period and frequency of the rotation, respectively. The angle swept by $\mathbf{r}(t)$ in t is therefore ωt (see Figure 3.1) for constant ω.*

*The quantity ω is $\dfrac{d\theta}{dt}$, the rate at which angle, θ, is swept by $\mathbf{r}$. If $\omega(t)$ is not a constant then the angle swept between times t_1 and t_2 is given by the integral,

$$\theta = \int_{t_1}^{t_2} \frac{d\theta}{dt} dt = \int_{t_1}^{t_2} \omega(t) dt.$$

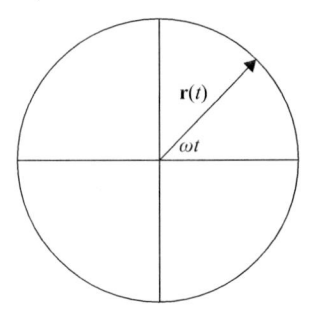

Fig. 3.1

Remember, the radian system of angular measurement is based on the idea that the circumference, C, of a circle of radius r is given by $C = 2\pi r$. If the circumference were broken up into q equal sections, the length of each one, called the arc length s would be given by

$$s = \frac{2\pi}{q} r, \tag{3.2}$$

where $q \times s = C$. The quantity s/r is therefore a measure of angle, which we say is in radians, and that means there are 2π radians in a complete rotation. Hence equation (3.1) gives the rate at which angle is swept by the radius vector rotating uniformly as just the total angle swept in one rotation divided by the time of the rotation. ω is measured in radians per second, rad/s.

If the radius vector $\mathbf{r}(t)$ represents the position of a body, then the speed of the body along the circular path is easily obtained as

$$v = \frac{2\pi r}{T} = \frac{2\pi}{T} r = \omega r. \tag{3.3}$$

Let us take a closer look at the rotating vectors in this motion. The velocity vector $\mathbf{v}(t)$ is always tangential to the circular path, and so it is a vector that is always 90° or $\pi/2$ radians in advance of the radius vector (see Figure 3.2).

Therefore $\mathbf{v}(t)$ is also a vector rotating at an angular frequency ω (Figure 3.2). The magnitude of $\mathbf{v}(t)$ is $v = \omega r$ as discussed above. It is important to realize that the two vector diagrams in Figure 3.2 are not in fact representing the same kinds of quantity; that is, one represents a rotating displacement vector of length r, whilst the other represents a rotating velocity vector of length $v = \omega r$.

The head of the vector $\mathbf{v}(t)$ describes a circle of circumference $2\pi v$ in a velocity coordinate system. The sum of all the little $\Delta \mathbf{v}$ around this path must therefore also have a length of $2\pi v$.* The rate of change $\mathbf{v}(t)$ is a vector, $\mathbf{a}(t)$ (see Figure 3.3), perpendicular to (90° or $\pi/2$ in advance of) $\mathbf{v}(t)$ with a magnitude

$$a = \frac{2\pi v}{T} = \frac{2\pi}{T} v = \omega v = \omega^2 r. \tag{3.4}$$

*Think about the line integrals introduced in the previous chapter: $\oint_{\text{circle}} dv = 2\pi v$.

56 Rotation

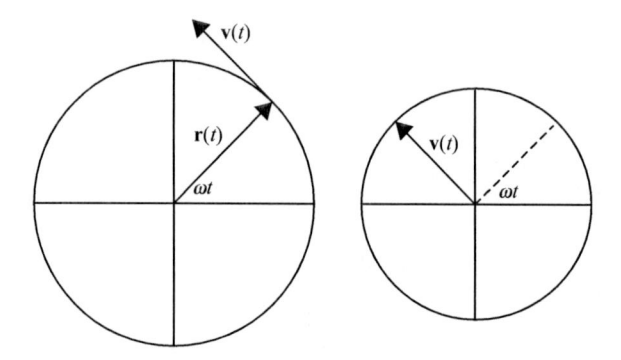

Fig. 3.2

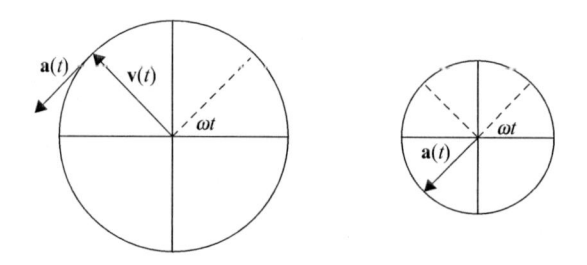

Fig. 3.3

The variable $a(t)$ is of course the *centripetal* acceleration vector. The vector diagrams in Figures 3.2 and 3.3 tell us that

$$\mathbf{a}(t) = -\omega^2 \mathbf{r}(t). \tag{3.5}$$

Expression (3.5) is a very important vector equation and it will pay dividends to analyse in more detail how it arises. Notice that $\mathbf{a}(t)$ itself is a rotating vector, antiparallel to the original radius vector $\mathbf{r}(t)$, hence the negative sign in (3.5).

Q1 Calculate how far away a spacecraft would have to be in order that the radius of the Earth's orbit would appear to be $1/3600°$ (one arc-second) wide.

Q2 A car has wheels of 40 cm in radius. It goes on a journey of exactly 140 km. Calculate: (1) how many times the wheels rotate on the journey and (2) the accuracy with which the radius would have to be known if your answer to part (1) should be correct to the nearest revolution. +

3.1.2 Workshop: Rotated coordinate systems and matrices

Imagine we have two sets of coordinate axes (x, y, z) and (x', y', z') with z and z' axes coincident and x, y, x', and y' axes rotated by some angle θ with respect to each other.

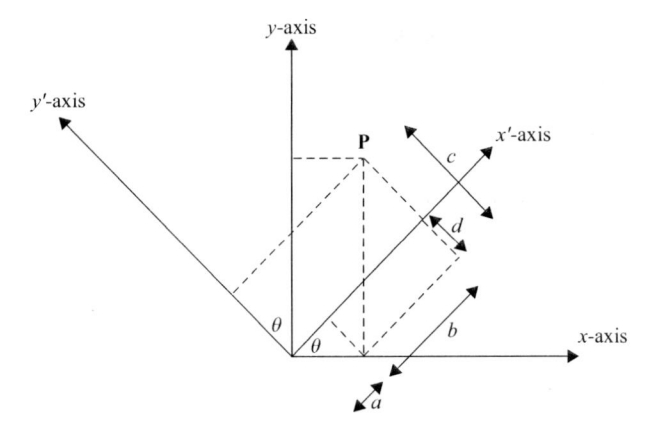

Fig. 3.4

The same point **P** is referred simultaneously to the two coordinate systems.

(a) By inspecting Figure 3.4 show that

$$a = x\cos(\theta) \quad b = y\sin(\theta) \quad c = y\cos(\theta) \quad d = x\sin(\theta). \tag{3.6}$$

So,

$$\begin{aligned} x' &= a + b \\ y' &= c - d. \end{aligned} \tag{3.7}$$

(b) Hence show that the transformation of coordinates may be written as matrices:

$$\begin{pmatrix} x' \\ y' \\ z' \end{pmatrix} = \begin{pmatrix} \cos\theta & \sin\theta & 0 \\ -\sin\theta & \cos\theta & 0 \\ 0 & 0 & 1 \end{pmatrix} \begin{pmatrix} x \\ y \\ z \end{pmatrix}. \tag{3.8}$$

Rotation transformations such as these are said to be *orthogonal*, so their matrix representations are *orthogonal matrices*. This means that the inverse (denoted with -1 at the top right) of the matrix is equal to the transpose (denoted with a T at the top right), which is obtained by swapping rows for columns and vice versa:

$$\begin{pmatrix} \cos\theta & \sin\theta & 0 \\ -\sin\theta & \cos\theta & 0 \\ 0 & 0 & 1 \end{pmatrix}^{-1} = \begin{pmatrix} \cos\theta & \sin\theta & 0 \\ -\sin\theta & \cos\theta & 0 \\ 0 & 0 & 1 \end{pmatrix}^{T} = \begin{pmatrix} \cos\theta & -\sin\theta & 0 \\ \sin\theta & \cos\theta & 0 \\ 0 & 0 & 1 \end{pmatrix}. \tag{3.9}$$

So,

$$\begin{pmatrix} x \\ y \\ z \end{pmatrix} = \begin{pmatrix} \cos\theta & -\sin\theta & 0 \\ \sin\theta & \cos\theta & 0 \\ 0 & 0 & 1 \end{pmatrix} \begin{pmatrix} x' \\ y' \\ z' \end{pmatrix}. \tag{3.10}$$

3.1.3 Workshop: Rotating vectors and the vector product

Using the results in the previous workshops, it is possible to obtain the rates of change of unit vectors rotating with respect to a fixed coordinate system. Let us begin with the transformation from the primed coordinates to the unprimed coordinates.

$$\begin{pmatrix} x \\ y \\ z \end{pmatrix} = \begin{pmatrix} \cos\theta & -\sin\theta & 0 \\ \sin\theta & \cos\theta & 0 \\ 0 & 0 & 1 \end{pmatrix} \begin{pmatrix} x' \\ y' \\ z' \end{pmatrix}. \tag{3.11}$$

If the primed coordinate frame is *rotating* at a uniform angular speed of ω about the z-axis, then $\theta(t) = \omega t$ so:

$$\begin{pmatrix} x \\ y \\ z \end{pmatrix} = \begin{pmatrix} \cos\omega t & -\sin\omega t & 0 \\ \sin\omega t & \cos\omega t & 0 \\ 0 & 0 & 1 \end{pmatrix} \begin{pmatrix} x' \\ y' \\ z' \end{pmatrix}. \tag{3.12}$$

(a) Differentiate (3.12) with respect to t (applying the product rule) and show that

$$\frac{d}{dt}\begin{pmatrix} x \\ y \\ z \end{pmatrix} = \omega \begin{pmatrix} -\sin\omega t & -\cos\omega t & 0 \\ \cos\omega t & -\sin\omega t & 0 \\ 0 & 0 & 0 \end{pmatrix} \begin{pmatrix} x' \\ y' \\ z' \end{pmatrix} + \begin{pmatrix} \cos\omega t & -\sin\omega t & 0 \\ \sin\omega t & \cos\omega t & 0 \\ 0 & 0 & 1 \end{pmatrix} \begin{pmatrix} \dot{x}' \\ \dot{y}' \\ \dot{z}' \end{pmatrix}, \tag{3.13}$$

where $\dot{x}'$ etc. are the time derivatives of positions in the primed coordinates.[*]

The second term in (3.13) is obviously the velocity vector in the primed coordinates as seen from the unprimed coordinates. If **P** were moving in the rotating coordinate system (primed system), this term would be non-zero.

The first term in (3.13) seems to have some interesting properties.

(b) Show that

$$\omega \begin{pmatrix} -\sin\omega t & -\cos\omega t & 0 \\ \cos\omega t & -\sin\omega t & 0 \\ 0 & 0 & 0 \end{pmatrix} \begin{pmatrix} x' \\ y' \\ z' \end{pmatrix} = \omega \begin{pmatrix} -y \\ x \\ 0 \end{pmatrix} \tag{3.14}$$

The following are the rules for multiplying unit vectors $\hat{\mathbf{x}}$, $\hat{\mathbf{y}}$, and $\hat{\mathbf{z}}$ by what mathematicians call the *vector product*.[†]

$$\begin{aligned} \hat{\mathbf{x}} \times \hat{\mathbf{x}} &= \hat{\mathbf{y}} \times \hat{\mathbf{y}} = \hat{\mathbf{z}} \times \hat{\mathbf{z}} = 0 \\ \hat{\mathbf{x}} \times \hat{\mathbf{y}} &= -\hat{\mathbf{y}} \times \hat{\mathbf{x}} = \hat{\mathbf{z}} \\ \hat{\mathbf{y}} \times \hat{\mathbf{z}} &= -\hat{\mathbf{z}} \times \hat{\mathbf{y}} = \hat{\mathbf{x}} \\ \hat{\mathbf{z}} \times \hat{\mathbf{x}} &= -\hat{\mathbf{x}} \times \hat{\mathbf{z}} = \hat{\mathbf{y}}. \end{aligned} \tag{3.15}$$

---

[*]Here we introduce the 'dot' notation to save room. We represent df/dt as $\dot{f}$ and d^2f/dt^2 as $\ddot{f}$, so the 'dots' keep track of the number of differentiations with respect to time.

[†]The other product that is useful to us in this text is the *scalar product*. As is obvious from their names, the scalar product produces a scalar and the vector product produces a vector. The scalar product was invaluable in considering *work* and *energy* in Chapter 2.

Notice that the vector product combines two unit vectors to form another vector (either $\mathbf{0}$, $\hat{\mathbf{x}}$, $\hat{\mathbf{y}}$, or $\hat{\mathbf{z}}$). The $\mathbf{0}$ is actually a vector with zero components.

(c) Using the rules for vector products of the unit vectors show that

$$\boldsymbol{\omega} \times \mathbf{r} = -\omega y \hat{\mathbf{x}} + \omega x \hat{\mathbf{y}} = \omega \begin{pmatrix} -y \\ x \\ 0 \end{pmatrix}, \tag{3.16}$$

where $\boldsymbol{\omega} = \omega \hat{\mathbf{z}}$ and $\mathbf{r} = x\hat{\mathbf{x}} + y\hat{\mathbf{y}} + z\hat{\mathbf{z}}$.

The vector $\boldsymbol{\omega}$ is called the *angular velocity* and is a vector of magnitude ω (the rate of rotation about the z-axis) and direction parallel to $\hat{\mathbf{z}}$, the unit vector in the z-direction. The vector $\mathbf{r}$ is the position vector of $\mathbf{P}$ in the unprimed coordinate system.

Combing our results from (a)–(c) this means our time derivative of the vector $\mathbf{r}$ could be written symbolically:

$$\frac{d_a}{dt}\mathbf{r} = \frac{d_r}{dt}\mathbf{r} + \boldsymbol{\omega} \times \mathbf{r}, \tag{3.17}$$

where $(d_a/dt)\mathbf{r}$ is the absolute velocity of $\mathbf{P}$ in the unprimed Galilean (non-rotating) coordinate system, and $(d_r/dt)\mathbf{r}$ is the velocity of $\mathbf{P}$ relative to the primed rotating coordinate system.

3.1.4 Angular velocity

The results of the last two workshops are really very important. The analyses yield the idea that the time derivatives of position vectors referred to two coordinate systems, primed (rotating) and unprimed (Galilean), are related thus:

$$\frac{d_a}{dt}\mathbf{r} = \frac{d_r}{dt}\mathbf{r} + \boldsymbol{\omega} \times \mathbf{r}. \tag{3.18}$$

Let us take a closer look at the term containing the vector product. To do this we make $\mathbf{P}$ stationary with respect to the primed coordinate system so (3.18) becomes

$$\frac{d_a}{dt}\mathbf{r} = \boldsymbol{\omega} \times \mathbf{r}. \tag{3.19}$$

Figure 3.5 shows a pictorial representation of expression (3.19). Point $\mathbf{P}$ rotates about the z-axis and its instantaneous position is given by the trio of numbers (x, y, z). As can be seen from the figure, in all of the following reasoning we are assuming that the origin of coordinates (from which $\mathbf{r}$ is measured) is on the axis of rotation.

Let us imagine a clock face, with the minute hand rotating clockwise. What direction would one associate with this motion? Up towards 12 o'clock because the hand sometimes points that way? Towards 3 o'clock because the hand sometimes points that way? Both are equally unhelpful. In fact the only way of choosing a direction that will always apply is to assign the rotation 'direction' *perpendicular* to the clock face – the direction in which the hands *never* point.

Point $\mathbf{P}$ is rotating as shown in Figure 3.5 and the convention is to represent the 'direction' of this rotation parallel to the positive z-axis (parallel to $+\hat{\mathbf{z}}$). Various

60 *Rotation*

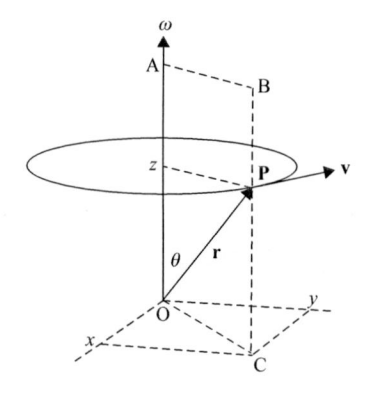

Fig. 3.5

aides-memoire have been created to help physicists remember this convention, but the most obvious one is to consider a right-hand screw. When turned in the way **P** is rotating about the z-axis, a right-hand screw will move upwards in the direction of the positive z-axis. Notice that the rotation in the opposite direction would be represented by a vector in the $-\hat{\mathbf{z}}$ direction.

With this convention established, we can see that the rotation of **P** about the z-axis can be described by an *angular velocity* vector $\boldsymbol{\omega} = \omega\hat{\mathbf{z}}$. Looking at Figure 3.5, we can immediately see that

$$\omega r \sin\theta = v, \qquad (3.20)$$

where r and v are, respectively, the magnitudes of **r** and **v**, and θ is the angle between the vectors $\boldsymbol{\omega}$ and **r**. The direction of **v** is *perpendicular* to the plane OABC, which is a plane that contains both $\boldsymbol{\omega}$ and **r**. So (3.20) may be recast as

$$|\boldsymbol{\omega}||\mathbf{r}|\sin\theta\,\hat{\mathbf{n}} = \mathbf{v}, \qquad (3.21)$$

where $\hat{\mathbf{n}}$ is a unit vector that is perpendicular to the plane OABC in the direction of **v** (see Figure 3.5). The vector, **v**, is effectively constructed out of the two vectors ω and **r**. Indeed, mathematicians would say that **v** is the *vector product* of $\boldsymbol{\omega}$ and **r**, written as

$$\boldsymbol{\omega} \times \mathbf{r} = |\boldsymbol{\omega}||\mathbf{r}|\sin\theta\,\hat{\mathbf{n}} = \mathbf{v}. \qquad (3.22)$$

Now, we already have an expression for $\boldsymbol{\omega} \times \mathbf{r}$ from Section 3.1.3:

$$\boldsymbol{\omega} \times \mathbf{r} = -\omega y\hat{\mathbf{x}} + \omega x\hat{\mathbf{y}} = \omega \begin{pmatrix} -y \\ x \\ 0 \end{pmatrix}, \qquad (3.23)$$

and we can now see the meaning of the swapping and change of sign of the components with respect to $\mathbf{r} = x\hat{\mathbf{x}} + y\hat{\mathbf{y}} + z\hat{\mathbf{z}}$. The vector $\boldsymbol{\omega} \times \mathbf{r}$ sits parallel to the x–y plane, perpendicular to the plane OABC, with a magnitude:

$$\omega\sqrt{x^2 + y^2} = |\boldsymbol{\omega}||\mathbf{r}|\sin\theta; \qquad (3.24)$$

that is,
$$\mathbf{v} = \boldsymbol{\omega} \times \mathbf{r} = -\omega y \hat{\mathbf{x}} + \omega x \hat{\mathbf{y}} = \omega \begin{pmatrix} -y \\ x \\ 0 \end{pmatrix}. \qquad (3.25)$$

We have effectively recovered our expression (3.3), $v = \omega r$, for circular motion, but it is now in a more generalized vector form: $\mathbf{v} = \boldsymbol{\omega} \times \mathbf{r}$. We have achieved this by inventing the angular velocity vector, $\boldsymbol{\omega}$, which represents the magnitude (rate) of rotation as well as the direction of rotation.

One final thing, as with the aides-memoire for remembering the direction of $\boldsymbol{\omega}$, there is a way of easily deducing the direction of a vector product. In performing the vector product $\mathbf{v} = \boldsymbol{\omega} \times \mathbf{r}$, one may think of a *right-hand screw* rotation of $\boldsymbol{\omega}$ into $\mathbf{r}$; that is, the direction of $\mathbf{v}$. Therefore the vector product $\mathbf{r} \times \boldsymbol{\omega}$ would be in the opposite direction.

Q3 Some chewing gum has become stuck on the tread of a tyre. Calculate the speed of the gum with respect to the axle if it is 40 cm from the axle, which is attached to a car travelling at 30 m/s. Assume no slip.

Q4 Repeat Q3 for the instantaneous velocity of the chewing gum with respect to the point of contact with the ground as a function of its height above the ground. +

Q5 An object is being rotated about the z-axis with angular frequency 10 rad/s. What is the velocity of the point which is momentarily at position (5 cm, 0, 3 cm)? Where will this point be 5 s later? +

Q6 An object is being rotated about an axis which passes the origin of coordinates. One part of the object is at position (x, y, z) when its velocity is (u, v, w). Give the angular velocity $\boldsymbol{\omega}$ as a vector, and explain how you would test whether the motion can be described as a pure rotation. ++

3.1.5 Workshop: Vector triple product

In the workshops and sections that follow, we will be seeing vectors formed by taking products like

$$\mathbf{a} \times (\mathbf{b} \times \mathbf{c}) \quad \text{and} \quad (\mathbf{a} \times \mathbf{b}) \times \mathbf{c}, \qquad (3.26)$$

where $\mathbf{a}$, $\mathbf{b}$, and $\mathbf{c}$ are arbitrary vectors. These products are called *vector triple products*. Let us consider $\mathbf{a} \times (\mathbf{b} \times \mathbf{c})$ first.

(a) Using the rules for vector products of unit vectors (3.1.3) show that

$$\begin{aligned}(\mathbf{b} \times \mathbf{c}) &= (b_x \hat{\mathbf{x}} + b_y \hat{\mathbf{y}} + b_z \hat{\mathbf{z}}) \times (c_x \hat{\mathbf{x}} + c_y \hat{\mathbf{y}} + c_z \hat{\mathbf{z}}) \\ &= \hat{\mathbf{x}}(b_y c_z - b_z c_y) + \hat{\mathbf{y}}(b_z c_x - b_x c_z) + \hat{\mathbf{z}}(b_x c_y - b_y c_x).^*\end{aligned} \qquad (3.27)$$

---

*A useful way of representing this result is by writing it in determinant form thus: $\mathbf{b} \times \mathbf{c} =$
$$\begin{vmatrix} \hat{\mathbf{x}} & \hat{\mathbf{y}} & \hat{\mathbf{z}} \\ b_x & b_y & b_z \\ c_x & c_y & c_z \end{vmatrix} = \hat{\mathbf{x}}(b_y c_z - b_z c_y) + \hat{\mathbf{y}}(b_z c_x - b_x c_z) + \hat{\mathbf{z}}(b_x c_y - b_y c_x).$$

62 *Rotation*

(b) Extend what you have done in (a) to show that

$$\mathbf{a} \times (\mathbf{b} \times \mathbf{c})$$
$$= \hat{\mathbf{x}}\{b_x(a_xc_x + a_yc_y + a_zc_z) - c_x(a_xb_x + a_yb_y + a_zb_z)\}$$
$$+ \hat{\mathbf{y}}\{b_y(a_xc_x + a_yc_y + a_zc_z) - c_y(a_xb_x + a_yb_y + a_zb_z)\}$$
$$+ \hat{\mathbf{z}}\{b_z(a_xc_x + a_yc_y + a_zc_z) - c_z(a_xb_x + a_yb_y + a_zb_z)\}. \quad (3.28)$$

(c) In Section 2.3, we introduced the *scalar product*. Collect together terms in the above expressions and use the rules for scalar products to show that

$$\mathbf{a} \times (\mathbf{b} \times \mathbf{c}) = (\mathbf{a} \cdot \mathbf{c})\mathbf{b} - (\mathbf{a} \cdot \mathbf{b})\mathbf{c}. \quad (3.29)$$

(d) Now show that

$$(\mathbf{a} \times \mathbf{b}) \times \mathbf{c} = (\mathbf{c} \cdot \mathbf{a})\mathbf{b} - (\mathbf{c} \cdot \mathbf{b})\mathbf{a}. \quad (3.30)$$

This would suggest that $\mathbf{a} \times (\mathbf{b} \times \mathbf{c})$ is *not* equal to $(\mathbf{a} \times \mathbf{b}) \times \mathbf{c}$ either in magnitude or direction.

3.1.6 Acceleration vectors in rotating frames

In part (a) of 3.1.3, we mention that the second term of our vector equation is obviously the velocity vector in the primed coordinates as seen from the unprimed coordinates. This means that if the two frames were not rotating with respect to each other and were merely *rotated* by an angle θ then

$$\begin{pmatrix} \dot{x} \\ \dot{y} \\ \dot{z} \end{pmatrix} = \begin{pmatrix} \cos\theta & -\sin\theta & 0 \\ \sin\theta & \cos\theta & 0 \\ 0 & 0 & 1 \end{pmatrix} \begin{pmatrix} \dot{x}' \\ \dot{y}' \\ \dot{z}' \end{pmatrix}; \quad (3.31)$$

that is, the rotation matrix used to transform the position vector for point **P** from unprimed to primed coordinates would work just as well for velocity vectors and indeed for any vectors. This is not hard to see. A vector is a quantity that has components in the x-, y-, and z-directions. Whilst these components do not necessarily have the meaning of *distances* along these axes, the magnitudes of the components may be represented as such (all that is needed is an appropriate scale). Therefore, since we derived our time-derivative expression using (3.31), we may think of differentiation of a vector **K** in a rotating coordinate system as applying the symbolic operation:

$$\frac{d_a}{dt} = \frac{d_r}{dt} + \boldsymbol{\omega} \times; \quad (3.32)$$

that is,

$$\frac{d_a}{dt}\mathbf{K} = \frac{d_r}{dt}\mathbf{K} + \boldsymbol{\omega} \times \mathbf{K}, \quad (3.33)$$

where $\boldsymbol{\omega}$ is the angular velocity of the rotating frame.

Let us apply (3.32) to the velocity vector $\mathbf{v}$:

$$\frac{d_a}{dt}\mathbf{v} = \frac{d_r}{dt}\mathbf{v} + \boldsymbol{\omega} \times \mathbf{v}, \qquad (3.34)$$

but

$$\mathbf{v} = \frac{d_a}{dt}\mathbf{r} = \frac{d_r}{dt}\mathbf{r} + \boldsymbol{\omega} \times \mathbf{r}. \qquad (3.35)$$

So (3.34) becomes

$$\frac{d_a}{dt}\mathbf{v} = \frac{d_r}{dt}\left(\frac{d_r}{dt}\mathbf{r} + \boldsymbol{\omega} \times \mathbf{r}\right) + \boldsymbol{\omega} \times \left(\frac{d_r}{dt}\mathbf{r} + \boldsymbol{\omega} \times \mathbf{r}\right). \qquad (3.36)$$

Expression (3.36) contains two things that might be new to the reader:

(i) The derivative of a vector product: $\dfrac{d_r}{dt}(\boldsymbol{\omega} \times \mathbf{r})$

(ii) The vector triple product: $\boldsymbol{\omega} \times (\boldsymbol{\omega} \times \mathbf{r})$.

To carry out (i), we need only respect the order of the vector product (remember that under the rules of vector products of unit vectors in Section 3.1.3, $\boldsymbol{\omega} \times \mathbf{r} \neq \mathbf{r} \times \boldsymbol{\omega}$). So,

$$\frac{d_r}{dt}(\boldsymbol{\omega} \times \mathbf{r}) = \frac{d_r}{dt}\boldsymbol{\omega} \times \mathbf{r} + \boldsymbol{\omega} \times \frac{d_r}{dt}\mathbf{r}. \qquad (3.37)$$

To carry out (ii) you will need to follow the rules derived in Section 3.1.5:

$$\boldsymbol{\omega} \times (\boldsymbol{\omega} \times \mathbf{r}) = (\boldsymbol{\omega} \cdot \mathbf{r})\boldsymbol{\omega} - (\boldsymbol{\omega} \cdot \boldsymbol{\omega})\mathbf{r}. \qquad (3.38)$$

Therefore (3.36) becomes

$$\frac{d_a}{dt}\mathbf{v} = \frac{d_r^2}{dt^2}\mathbf{r} + \frac{d_r}{dt}\boldsymbol{\omega} \times \mathbf{r} + 2\boldsymbol{\omega} \times \frac{d_r}{dt}\mathbf{r} + \boldsymbol{\omega} \times (\boldsymbol{\omega} \times \mathbf{r}). \qquad (3.39)$$

Here we have not expanded out the vector triple product.

Q7 Expression (3.39) may be derived by differentiating with respect to t a second time the expression (3.13) in Section 3.1.3. This exercise is likely to require a page of algebra and matrices. Notice that we needed to be careful with our notation and introduced $\dfrac{d_r}{dt}\mathbf{r}$ to mean:

$$\frac{d_r}{dt}\mathbf{r} = \begin{pmatrix} \cos\theta & -\sin\theta & 0 \\ \sin\theta & \cos\theta & 0 \\ 0 & 0 & 1 \end{pmatrix} \begin{pmatrix} \dot{x}' \\ \dot{y}' \\ \dot{z}' \end{pmatrix};$$

that is, the velocity relative to the *rotating* coordinate system. When differentiating a second time, you will need to remember this so that terms containing $\dot{x}'$, $\dot{y}'$, and $\dot{z}'$, or $\ddot{x}'$, $\ddot{y}'$, and $\ddot{z}'$, do in fact, respectively, involve the velocity in the rotating frame, or the acceleration in the rotating frame.

64   *Rotation*

Q8 A millstone with 60 cm radius is being rotated about the z-axis to grind some flour. Calculate the acceleration of a point on the rim of the millstone:
(a) when the millstone is rotating at a steady 8 rad/s
(b) just as the millstone begins slowing at a rate of 0.2 rad/s^2 from its original 8 rad/s.

Q9 Use (3.39) and the results from the workshop in Section 3.1.5 to consider the uniform circular motion of a point that is stationary in the primed coordinate frame (x', y', z') with respect to the unprimed coordinate frame (x, y, z). You should of course recover:

$$\frac{d_a}{dt}\mathbf{v} = \boldsymbol{\omega} \times (\boldsymbol{\omega} \times \mathbf{r}) = -\omega^2 \mathbf{r}.$$

3.1.7   'Fictitious force': Centrifugal and Coriolis forces

Let us look more closely at the meaning of (3.39). To do this, we will consider the experiences of astronauts living in a giant rotating space station. The living quarters of this station are on the rim of a huge wheel and it is the rotation that provides the artificial gravity.

The coordinate system attached to the space station is the primed coordinate system so setting $(d_r/dt)\boldsymbol{\omega} = \mathbf{0}$ (as the rate of rotation is not increasing) in (3.39) would give

$$\frac{d_a}{dt}\mathbf{v} = \frac{d_r^2}{dt^2}\mathbf{r} + 2\boldsymbol{\omega} \times \frac{d_r}{dt}\mathbf{r} + \boldsymbol{\omega} \times (\boldsymbol{\omega} \times \mathbf{r}). \tag{3.40}$$

Let us see what an astronaut standing on the rim of the rotating space station experiences first. We rearrange (3.40) so that we have the acceleration as measured by the astronaut, $(d_r^2/dt^2)\mathbf{r}$:

$$\frac{d_r^2}{dt^2}\mathbf{r} = \frac{d_a}{dt}\mathbf{v} - 2\boldsymbol{\omega} \times \frac{d_r}{dt}\mathbf{r} - \boldsymbol{\omega} \times (\boldsymbol{\omega} \times \mathbf{r}). \tag{3.41}$$

Since he is in equilibrium and static with respect to the space station (the rotating coordinate system), his $(d_r/dt)\mathbf{r} = \mathbf{0}$ and he experiences no resultant force. Newton's second law applied to this astronaut would give us

$$\mathbf{0} = \mathbf{F} - m\boldsymbol{\omega} \times (\boldsymbol{\omega} \times \mathbf{r}), \tag{3.42}$$

where m is the mass of the astronaut. Looking at the direction* of the force vector $-m\boldsymbol{\omega} \times (\boldsymbol{\omega} \times \mathbf{r})$, we see that $\mathbf{F}$ must be directed towards the centre of the space station. This is of course what we would expect to be the force on the astronaut referred to the absolute Galilean (unprimed) frame – this is the *centripetal force* (centre-seeking).

---

*For the direction of this vector triple product, work out the direction of $\boldsymbol{\omega} \times \mathbf{r}$ first using the right-hand screw rule (as suggested in Section 3.1.4), then do exactly the same to do the vector product of $\boldsymbol{\omega}$ with the answer to your first $\boldsymbol{\omega} \times \mathbf{r}$.

Rotational kinematics and dynamics 3.1

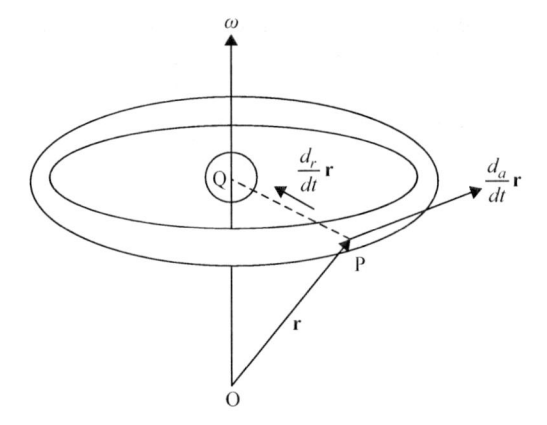

Fig. 3.6

The force $-m\boldsymbol{\omega} \times (\boldsymbol{\omega} \times \mathbf{r})$ is therefore the *centrifugal force*, and whilst it is often called a fictitious force it is very real to the astronaut.

Now let us turn our attention to an astronaut moving up a service ladder towards the centre of the space station. The astronaut moves at a velocity $(d_r/dt)\mathbf{r}$ with respect to the space station (in Figure 3.6, this vector is labelled $(d_r/dt)\mathbf{r}$).

According to (3.41), he should experience a force

$$-2m\boldsymbol{\omega} \times \frac{d_r}{dt}\mathbf{r}. \tag{3.43}$$

To see this remember that the vector $\boldsymbol{\omega}$ is directed as shown in Figure 3.6, so the cross product $-\boldsymbol{\omega} \times (d_r/dt)\mathbf{r}$ will be a vector directed in the same direction as $(d_a/dt)\mathbf{r}$. To see how this apparent force arises, we must remember that as the astronaut moves up the ladder he is moving into parts of the space station that are in fact moving more slowly than he is. To the astronaut therefore there seems to be some mysterious force that is pushing him against the ladder as he climbs up it – we know though that the normal reaction from the ladder provides the force required to slow him down so that he is moving at the same speed as that part of the space station. The force, $-2m\boldsymbol{\omega} \times (d_r/dt)\mathbf{r}$, is called the *Coriolis* force after the French mathematician Gaspard Gustave de Coriolis (1792–1843).

Q10 Use the analysis that led to (3.39) to analyse the motion of a particle moving at a constant velocity along the x-axis of the unprimed coordinate frame (the Galilean frame). You will discover that whilst

$$\frac{d_a}{dt}\mathbf{v} = \frac{d_a^2}{dt^2}\mathbf{r} = \mathbf{0},$$

$(d_r^2/dt^2)\mathbf{r}$ will not be ZERO and will be made up of two components. The particle's motion with respect to the primed coordinate frame will be accelerating – the dynamics would therefore require that there be *fictitious forces* to explain why the particle in question does not follow a straight line in the primed coordinate frame. +

3.2 Orbits

3.2.1 The Kepler problem

Johannes Kepler's (1571–1630) work of 1609 was based on a meticulous analysis of the accurate astronomical data collected by Tycho Brahe (1546–1601). Working from a Copernican viewpoint, Kepler found that he was unable to fit the orbit of Mars onto a circle. Even the inclusion of epicycles (essentially orbits on orbits) could not model accurately the orbit of Mars. Kepler had great faith in the reliability of Brahe's data and continued to search for an explanation for the irregularities in the Martian orbit. Finally he was able to make the orbit of Mars fit a particular trajectory but only if he dropped an assumption that he shared with both Copernicus and the Greeks: that the 'natural' path of a planet was a circle. If the orbit of Mars were an ellipse with the Sun at one of the foci, then the Copernican model fits the data extremely well without the need for epicycles.

Kepler summarized his discoveries in three laws that now bear his name:

- *Kepler's first law*: Planets move on ellipses with the Sun at one of the foci (See Figure 3.7).
  - In more elementary work, we learn that an ellipse may be drawn by using a loop of inextensible string, two pins, and a pencil. The pins are placed at the foci ($\mathbf{F_1}$ and $\mathbf{F_2}$) and the pencil traces a curve keeping the string taut.
  - The distance $\mathbf{OP}$ is called the *semi-major axis* and is usually given the symbol a. The distance $\mathbf{OQ}$ is called the *semi-minor axis* and is usually given the symbol b.
  - $\mathbf{F_1}$ $\mathbf{R}$ $\mathbf{F_2}$ is a constant distance no matter where $\mathbf{R}$ is on the ellipse. The total distance $\mathbf{F_1}$ $\mathbf{R}$ $\mathbf{F_2}$ is equal to $2a$.
  - The distance $\mathbf{OF_1}$ (or $\mathbf{OF_2}$) is often expressed as a factor e times the semi-major axis a; that is, $\mathbf{OF_1} = \mathbf{OF_2} = ea$. The factor e is called the *eccentricity* and is about 0.81 in the ellipse of Figure 3.7.
  - Using triangle $\mathbf{OF_1Q}$ (or $\mathbf{OF_2Q}$), we see that a, b, and e are linked, thus

$$a = \frac{b}{\sqrt{1-e^2}}. \tag{3.44}$$

- *Kepler's second law*: The imaginary line connecting the planet to the Sun sweeps out equal areas over equal periods of time.
- *Kepler's third law*: The square of the orbital period (T) for a planet is proportional to the cube of the semi-major axis (a) of the ellipse; that is, $T^2 \propto a^3$.

Kepler's three laws describe, but they do not explain. By the time Newton came onto the scene in the late 1660s, the derivation of Kepler's three laws of planetary motion from yet more fundamental laws of nature became dubbed the Kepler problem.

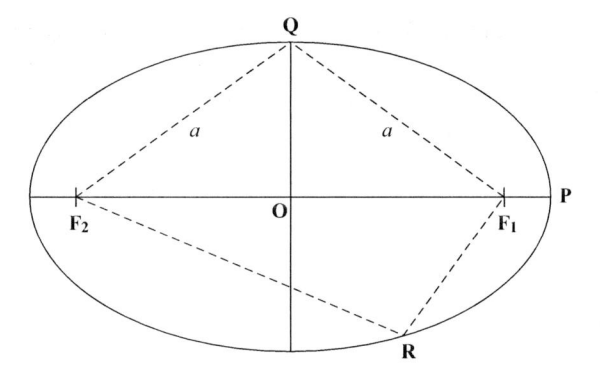

Fig. 3.7

It seems that Newton was aware that an inverse-square* force law could lead to closed trajectories that were conic sections. Conic sections are curves that are obtained from cutting (taking sections) through cones. Geometry text books are the places to go for a really in-depth study of such curves. Presently, we are only interested in the polar coordinate representation of these things. We include it in this text because this form for conic sections is particularly useful for the study of trajectories that result from an inverse-square force law.

Pappus of Alexandria (c. AD 300) came up with a way of describing the conic sections:

A conic section is made up of all the points **R** that are a distance r from a point **F** (which we call the focus) such that the point is also a distance r/e from a fixed line called the *directrix*, **D**. The constant e is the *eccentricity* of the conic section.

- For $0 < e < 1$, the conic section is an ellipse.
- For $e = 1$, the conic section is a parabola.
- For $e > 1$, the conic section is a hyperbola.

From Figure 3.8, we can see that $d = r \cos\theta + r/e$, which rearranges to

$$r(\theta) = \frac{ed}{1 + e \cos\theta}. \tag{3.45}$$

In Chapter 2, we introduced the concept of a gravitational field around a point mass. The gravitational field vector $\mathbf{g}_1$ at a distance r away from a mass m_1 is

$$\mathbf{g}_1 = -G\frac{m_1}{r^3}\mathbf{r}, \tag{3.46}$$

where **r** is the position vector of the point we are interested in relative to m_1.

*Early in his investigations, Newton realized that two special power laws enjoyed a degree of mathematical elegance not shared by the others: force decreasing as the square of the distance and force increasing directly as the distance. Only for these two laws are the orbits conic sections.

68 Rotation

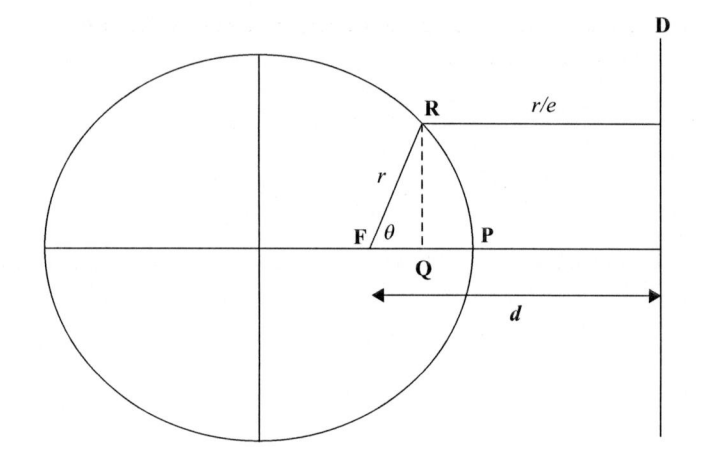

Fig. 3.8

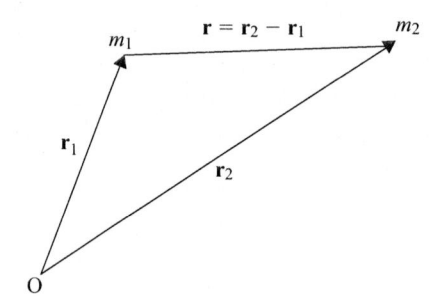

Fig. 3.9

The 'weight' of m_2 in the gravitational field of m_1 is therefore

$$\mathbf{W} = m_2 \mathbf{g}_1 = -G\frac{m_2 m_1}{r^3}\mathbf{r}. \tag{3.47}$$

We are now in a position to attack the Kepler problem. In Figure 3.9, two masses m_1 and m_2 have position vectors $\mathbf{r}_1$ and $\mathbf{r}_2$ in some coordinate system with origin O. The gravitational interaction between m_1 and m_2 leads to the following forces between them:

$$\begin{aligned}\text{Weight of } m_1 \text{ in the field of } m_2 &= \mathbf{W}_1 = G\frac{m_1 m_2}{r^3}\mathbf{r}, \\ \text{Weight of } m_2 \text{ in the field of } m_1 &= \mathbf{W}_2 = -G\frac{m_2 m_1}{r^3}\mathbf{r},\end{aligned} \tag{3.48}$$

where $\mathbf{r} = \mathbf{r}_2 - \mathbf{r}_1$.

Now $\mathbf{W}_1$ is the resultant force acting on m_1 and $\mathbf{W}_2$ is the resultant force acting on m_2, so Newton's second law would give us

$$m_1 \frac{d^2}{dt^2} \mathbf{r}_1 = \mathbf{W}_1 = G \frac{m_1 m_2}{r^3} \mathbf{r},$$
$$m_2 \frac{d^2}{dt^2} \mathbf{r}_2 = \mathbf{W}_2 = -G \frac{m_2 m_1}{r^3} \mathbf{r}. \tag{3.49}$$

These two expressions must necessarily sum to zero (Newton's third law):

$$m_1 \frac{d^2}{dt^2} \mathbf{r}_1 + m_2 \frac{d^2}{dt^2} \mathbf{r}_2 = \mathbf{0}. \tag{3.50}$$

Integration of (3.50) leads to

$$m_1 \frac{d}{dt} \mathbf{r}_1 + m_2 \frac{d}{dt} \mathbf{r}_2 = \mathbf{p}, \tag{3.51}$$
$$m_1 \mathbf{r}_1 + m_2 \mathbf{r}_2 = \mathbf{p}t + \mathbf{q}, \tag{3.52}$$

where $\mathbf{p}$ is a constant momentum vector and $\mathbf{q}$ is a constant displacement vector (multiplied by the total mass). Expression (3.51) is a statement of the conservation of linear momentum and (3.52) describes how the centre of mass drifts with respect to the origin of our coordinate system O. The coordinates of the centre of mass may be obtained by finding the displacement vector $\mathbf{R}$ of a mass $(m_1 + m_2)$ such that

$$m_1 \mathbf{r}_1 + m_2 \mathbf{r}_2 = (m_1 + m_2)\mathbf{R}. \tag{3.53}$$

Combining the expressions in (3.49), we obtain

$$\frac{d^2}{dt^2} \mathbf{r} = -G \frac{(m_1 + m_2)}{r^3} \mathbf{r}, \tag{3.54}$$

which is a differential equation in $\mathbf{r} = r\hat{\mathbf{x}}'$ (Figure 3.10). The solution of this would give the trajectories of the mass m_2 in the coordinate system centred on m_1.

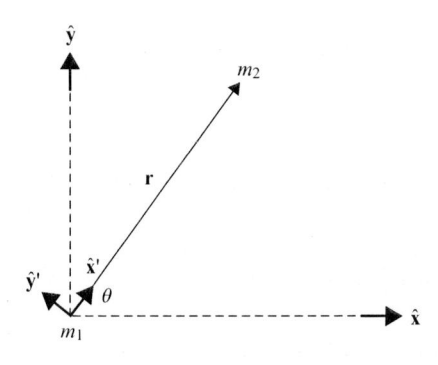

Fig. 3.10

70   Rotation

Q11 A satellite is in a circular orbit of radius r about a spherical planet of mass M. Show that the orbital period for the satellite is given by

$$T = \frac{2\pi}{\sqrt{GM}} r^{\frac{3}{2}}.$$

Q12 If the above satellite is in a low altitude orbit, show that the period is determined entirely by the average density of the planet and not at all by the size of the planet. +

Q13 A binary star system made up of two stars of mass m_1 and m_2 in circular motion about their centre of mass and separated by a distance a. Show that

$$T^2 = \frac{4\pi^2}{G(m_1 + m_2)} a^3.$$

Compare this expression with the more general expression (3.80) in Section 3.2.4.

3.2.2   Kepler's first law and properties of $\frac{d^2}{dt^2}\mathbf{r}$

The acceleration in (3.54) has some interesting properties because its direction is always antiparallel to the vector $\mathbf{r}$. We can probe some of these properties by constructing various products* of $(d^2/dt^2)\mathbf{r}$ with other vectors in the system. To be more economical on the notation let

$$\frac{d^2}{dt^2}\mathbf{r} = \mathbf{a} \quad \text{and} \quad \frac{d}{dt}\mathbf{r} = \mathbf{v}.$$

Then,

$$\mathbf{r} \times \mathbf{a} = -\frac{G(m_1 + m_2)}{r^3} \mathbf{r} \times \mathbf{r} = \mathbf{0}, \tag{3.55}$$

and since the vector product of any vector with itself is zero we can write (3.55) as

$$\frac{d}{dt}(\mathbf{r} \times \mathbf{v}) = \mathbf{0}, \tag{3.56}$$

which implies that

$$\mathbf{r} \times \mathbf{v} = \text{a constant vector}. \tag{3.57}$$

This quantity, which we shall call $\mathbf{h}$ here, is a constant of the motion and is therefore one of the conserved quantities that we will find invaluable in orbital motion problems. In Section 3.2.3, the workshop uses (3.56) and (3.57) to understand the underlying reasons behind Kepler's second law.

When we take the scalar product of (3.54) with $\mathbf{v}$, we get

$$\mathbf{v} \cdot \mathbf{a} = -\frac{G(m_1 + m_2)}{r^3} \mathbf{v} \cdot \mathbf{r}. \tag{3.58}$$

*Either scalar, vector or vector triple products.

Orbits 3.2 71

The left-hand side of this is just the time derivative of $(v^2/2)$,* where v is the magnitude of the vector $\mathbf{v}$.

The right-hand side of (3.58) requires a little more manipulation,† but also turns out to be the time derivative of something quite familiar:

$$-\frac{G(m_1+m_2)}{r^3}\mathbf{v}\cdot\mathbf{r} = \frac{d}{dt}\left\{\frac{G(m_1+m_2)}{r}\right\}. \tag{3.59}$$

Therefore (3.58) becomes

$$\frac{d}{dt}\left\{\frac{v^2}{2} - \frac{G(m_1+m_2)}{r}\right\} = 0, \tag{3.60}$$

or

$$\frac{v^2}{2} - \frac{G(m_1+m_2)}{r} = \text{a constant scalar}. \tag{3.61}$$

This of course is another conserved quantity and the workshop in 7.11 takes (3.61) further and classifies orbits in terms of the values of this conserved quantity.

Finally, taking the vector product of $\mathbf{a}$ with $\mathbf{r} \times \mathbf{v}$ or $\mathbf{h}$ leads to something very interesting indeed:

First of all,

$$\mathbf{h} = \mathbf{r}\times\mathbf{v} = \mathbf{r}\times\left(\frac{d_r}{dt}\mathbf{r}+\boldsymbol{\omega}\times\mathbf{r}\right) = \mathbf{r}\times\left(\frac{dr}{dt}\hat{\mathbf{x}}'+\omega r\hat{\mathbf{y}}'\right) = \omega r^2\hat{\mathbf{z}}, \tag{3.62}$$

where we align the rotating x'-axis with the position vector $\mathbf{r}$, with $\hat{\mathbf{y}}'$ perpendicular to $\hat{\mathbf{x}}'$, but still in the plane of rotation (see Figure 3.10). So,

$$\mathbf{a}\times\mathbf{h} = -G\frac{(m_1+m_2)}{r^3}\mathbf{r}\times\omega r^2\hat{\mathbf{z}} = -G\frac{(m_1+m_2)}{r^3}r^3\omega\hat{\mathbf{x}}'\times\hat{\mathbf{z}}, \tag{3.63}$$

---

* $\dfrac{d}{dt}\left\{\dfrac{v^2}{2}\right\} = \dfrac{1}{2}\dfrac{d}{dt}(\mathbf{v})\cdot\mathbf{v}+\dfrac{1}{2}\mathbf{v}\cdot\dfrac{d}{dt}(\mathbf{v}) = \mathbf{v}\cdot\mathbf{a}$.

† $\dfrac{d}{dt}\left\{\dfrac{G(m_1+m_2)}{r}\right\}$ is not as simple as it looks as $r = \sqrt{x^2+y^2+z^2}$ when referred to Cartesian coordinates: $\dfrac{d}{dt}\left\{\dfrac{G(m_1+m_2)}{\sqrt{x^2+y^2+z^2}}\right\} = G(m_1+m_2)\left[-\dfrac{1}{2}\dfrac{1}{(x^2+y^2+z^2)^{\frac{3}{2}}}2x\dfrac{dx}{dt}+\cdots\right]$, with the expression in the square brackets containing a term for each of the three dimensions of space. This may be simplified to: $\dfrac{d}{dt}\left\{\dfrac{G(m_1+m_2)}{\sqrt{x^2+y^2+z^2}}\right\} = -\dfrac{G(m_1+m_2)}{r^3}\left[x\dfrac{dx}{dt}+y\dfrac{dy}{dt}+z\dfrac{dz}{dt}\right] = -\dfrac{G(m_1+m_2)}{r^3}\mathbf{r}\cdot\mathbf{v}$,
which is of course the right-hand side of (3.59).

72 Rotation

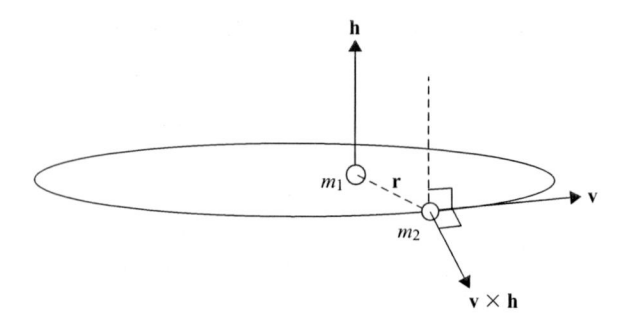

Fig. 3.11

which of course is just:

$$\mathbf{a} \times \mathbf{h} = G(m_1 + m_2)\omega\hat{\mathbf{y}}'. \tag{3.64}$$

But*

$$\omega\hat{\mathbf{y}}' = \frac{d\hat{\mathbf{x}}'}{dt}. \tag{3.65}$$

Once again we can rewrite (3.64) as the time derivative of something[†]:

$$\frac{d}{dt}(\mathbf{v} \times \mathbf{h}) = \frac{d}{dt}\left\{G(m_1 + m_2)\hat{\mathbf{x}}'\right\}, \tag{3.66}$$

which means that

$$\mathbf{v} \times \mathbf{h} - G(m_1 + m_2)\hat{\mathbf{x}}' = \text{a constant vector}. \tag{3.67}$$

As we can see from Figure 3.11, the vector $\mathbf{v} \times \mathbf{h}$ is always perpendicular to the tangent on the trajectory and is always pointing outwards away from the orbiting masses. We may choose a form for the constant vector so as to clarify its meaning. Remember, this vector has arisen out of the integration of a time derivative, so its value is set by the boundary conditions[‡] of the system.

In Figure 3.12, we see two possible values of the constant vector. On the left, we have the situation when the constant vector is zero, so the vector $\mathbf{v} \times \mathbf{h}$ is directed along the $\hat{\mathbf{x}}'$-axis. In the picture on the right, we see that the non-zero constant vector (which we have chosen to be parallel to the $\hat{\mathbf{x}}$-axis without loss of generality) shifts the circle on which the end of the vector $\mathbf{v} \times \mathbf{h}$ resides.

---

*$\dfrac{d_a \hat{\mathbf{x}}'}{dt} = \dfrac{d_r \hat{\mathbf{x}}'}{dt} + \boldsymbol{\omega} \times \hat{\mathbf{x}}' = \omega\hat{\mathbf{z}} \times \hat{\mathbf{x}}' = \omega\hat{\mathbf{y}}'$, as $\dfrac{d_r \hat{\mathbf{x}}'}{dt} = 0$ because $\hat{\mathbf{x}}'$ is stationary with respect to the rotating coordinate system.

[†] As $\mathbf{h}$ is itself a constant vector $\dfrac{d}{dt}(\mathbf{v} \times \mathbf{h}) = \mathbf{a} \times \mathbf{h}$.

[‡] When one solves a differential equation the resulting solution is of course a general solution; that is, it describes *all* the possible solutions to the equation. To make the solution specific to the physical system under study, one needs to introduce the *boundary conditions*, which are specific values of the variables at their extremities, for example at $t=0$ or $t=\infty$ (hence the word 'boundary').

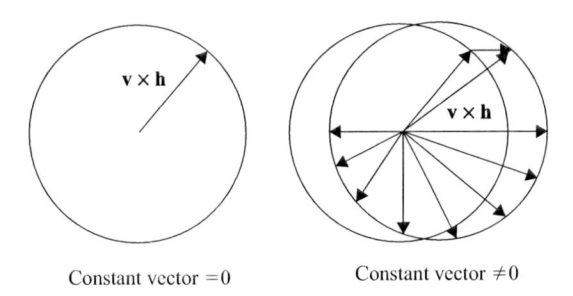

Fig. 3.12

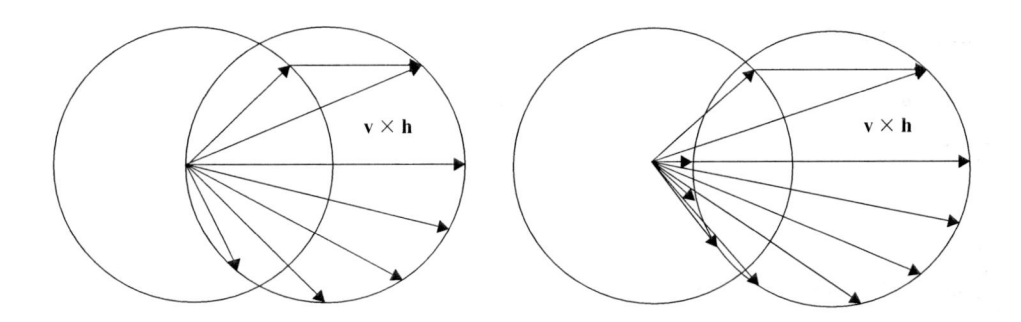

Fig. 3.13

The vector $\mathbf{v} \times \mathbf{h}$ sits in the same plane as $\mathbf{v}$. The situation on the left-hand side of Figure 3.12 is obviously the condition for a *circular* orbit as $\mathbf{v} \times \mathbf{h}$ is along $\hat{\mathbf{x}}'$ and therefore $\hat{\mathbf{x}}'$ and hence $\mathbf{r}$ is always perpendicular to the velocity $\mathbf{v}$. The situation on the right must therefore be the condition for an *elliptical* orbit with $\mathbf{v} \times \mathbf{h}$ having maximum and minimum magnitudes corresponding, respectively, to the points of minimum and maximum magnitudes of $\mathbf{r}$.

As we shift the two circles further apart (Figure 3.13) with a constant vector equal and exceeding in magnitude of $\mathbf{v} \times \mathbf{h}$, we see that $\mathbf{v} \times \mathbf{h}$ never takes values along the negative $\hat{\mathbf{x}}$-axis. Thus, the orbits corresponding to these choices of the constant vector cannot be periodic and these values of $\mathbf{v} \times \mathbf{h}$ turn out to be values taken up as the magnitude of $\mathbf{r}$ tends to infinity. In these two cases then we have respectively the *parabolic* and *hyperbolic* orbits.

This pictorial analysis of $\mathbf{v} \times \mathbf{h}$ means that we can integrate (3.66) to obtain

$$\mathbf{v} \times \mathbf{h} = G(m_1 + m_2)(\hat{\mathbf{x}}' + \mathbf{e}), \qquad (3.68)$$

where $G(m_1 + m_2)\mathbf{e}$ is the constant vector and $\mathbf{e}$ is a vector with a magnitude:

- $0 \leq e < 1$ for a periodic orbit (circle or ellipse)
- $e = 1$ for a parabolic orbit
- $e > 1$ for a hyperbolic orbit.

74 Rotation

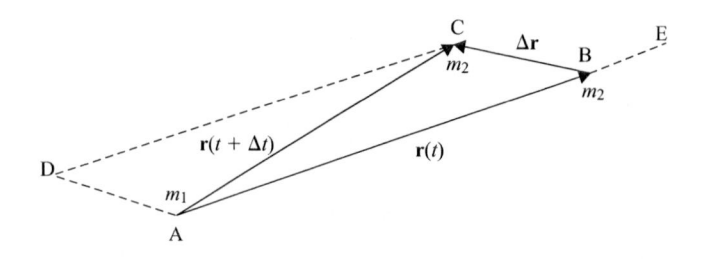

Fig. 3.14

We can now take the scalar product of **r** with **v** × **h** and obtain

$$\mathbf{r} \cdot (\mathbf{v} \times \mathbf{h}) = G(m_1 + m_2)(r + er\cos\theta), \tag{3.69}$$

where θ is the angle between **r** and **e**. Now* $\mathbf{r} \cdot (\mathbf{v} \times \mathbf{h}) = (\mathbf{r} \times \mathbf{v}) \cdot \mathbf{h} = \mathbf{h} \cdot \mathbf{h} = h^2$, so

$$r(\theta) = \frac{h^2/G(m_1+m_2)}{(1+e\cos\theta)}, \tag{3.70}$$

which of course *is* the form for a conic section (3.45).

3.2.3 Workshop: Kepler's second law

In the previous section, the constant vector $\mathbf{h}\,(\mathbf{r}\times\mathbf{v}=\omega r^2\hat{\mathbf{z}})$ figures a lot in the analysis.

What is this constant?

In Figure 3.14, the radius vector **r** changes by a vector $\Delta\mathbf{r}$ over an interval Δt. The *vector* product (see Section 3.1.3) of **r** and $\Delta\mathbf{r}$ is a vector perpendicular to the plane containing **r** and $\Delta\mathbf{r}$ and has a magnitude that is the area of the parallelogram ABCD. The vector product $\mathbf{r}\times\Delta\mathbf{r}$ is

$$|\mathbf{r}||\Delta\mathbf{r}|\sin\phi\,\hat{\mathbf{z}} = r\Delta r\sin\phi\,\hat{\mathbf{z}}, \tag{3.71}$$

where ϕ is the angle CBE in Figure 3.14.

(a) Show that $r\Delta r\sin\phi$ is just the area of the parallelogram ABCD.
(b) Hence show that the area swept, $\Delta\mathbf{A}$, by **r** in the time Δt is

$$\Delta\mathbf{A} = \frac{1}{2}\mathbf{r}\times\Delta\mathbf{r}. \tag{3.72}$$

In the limit as $\Delta t \to 0$, the instantaneous rate of change of this vector is given by

$$\frac{d}{dt}\mathbf{A} = \frac{1}{2}\mathbf{r}\times\frac{d}{dt}\mathbf{r}. \tag{3.73}$$

*$\mathbf{A}\cdot(\mathbf{B}\times\mathbf{C}) = \begin{vmatrix} A_x & A_y & A_z \\ B_x & B_y & B_z \\ C_x & C_y & C_z \end{vmatrix} = \begin{vmatrix} B_x & B_y & B_z \\ C_x & C_y & C_z \\ A_x & A_y & A_z \end{vmatrix} = \begin{vmatrix} C_x & C_y & C_z \\ A_x & A_y & A_z \\ B_x & B_y & B_z \end{vmatrix} = (\mathbf{A}\times\mathbf{B})\cdot\mathbf{C}$; that is,

(c) Hence show that

$$\frac{d}{dt}\mathbf{A} = \frac{1}{2}h\hat{z}. \tag{3.74}$$

Expression (3.74) essentially says that the rate at which area is swept by the radius vector is a constant of the motion. This is of course nothing more than Kepler's second law – the imaginary line connecting the planet to the Sun sweeps out equal areas over equal periods of time.

3.2.4 Workshop: Kepler's third law

We see from the previous section that the outcome of looking for an $r(\theta)$ is to rediscover the general form for a *conic section*. Remember, our analyses began by looking at the motion of two point masses about each other (m_1 and m_2) under the influence of the gravitational force. Let us just take stock of the major results:

(i) The general expression for a conic section in polar coordinates is

$$r(\theta) = \frac{p}{1 + e\cos\theta}, \tag{3.75}$$

where p is a constant and e is the *eccentricity*.
For
- $0 \leq e < 1$ the conic section is a circle or an ellipse.
- $e = 1$ the conic section is a parabola.
- $e > 1$ the conic section is a hyperbola.

(ii) The gravitational interaction between the two masses is through their centres, so provides no 'torque' (See Workshop 7.8). The consequence of this is that the radius vector **r** sweeps out equal areas in equal times:

$$\frac{d}{dt}\mathbf{A} = \frac{1}{2}h\hat{\mathbf{z}}. \tag{3.76}$$

Here **A** is a vector that has a magnitude (a constant h) which is the area swept per second by the radius vector and direction parallel to $\hat{\mathbf{z}}$.

(iii) The polar coordinate form of our solution for the trajectories is

$$r = \frac{h^2/G(m_1 + m_2)}{1 + e\cos(\theta)}. \tag{3.77}$$

(a) Taking the case of a closed orbit so that we can define a *time period* T, we have a general form for the trajectory in *Cartesian coordinates*:

$$\frac{x^2}{a^2} + \frac{y^2}{b^2} = 1, \tag{3.78}$$

76 Rotation

where a and b are related by equation (3.44) with $0 \leq e < 1$:

$$a = \frac{b}{\sqrt{1-e^2}},^*$$

show that

$$\frac{\pi ab}{T} = \frac{1}{2}\sqrt{pG(m_1+m_2)},^\dagger \qquad (3.79)$$

with

$$p = a(1-e^2).$$

(b) Hence show that

$$T^2 = \frac{4\pi^2}{G(m_1+m_2)}a^3. \qquad (3.80)$$

Q14 By using the expressions:

$$r = \frac{a(1-e^2)}{1+e\cos(\theta)}$$

and

$$h = \omega r^2,$$

which arose out of our analyses of planetary trajectories, show that
- $V^2 = G(m_1+m_2)\left(\frac{2}{r} - \frac{1}{a}\right)$ for an elliptical orbit,
- $V^2 = G(m_1+m_2)\left(\frac{2}{r}\right)$ for a parabolic orbit, and
- $V^2 = G(m_1+m_2)\left(\frac{2}{r} + \frac{1}{a}\right)$ for a hyperbolic orbit,

where V is the orbital speed: $V = \sqrt{\dot{r}^2 + \omega^2 r^2}$.

Q15 A spacecraft in an elliptical orbit around the Sun (mass M) has a period T. Rocket motors on-board the craft are fired and an impulse is imparted to the craft so that the orbital speed V is increased by an increment ΔV without a change in direction of the velocity. Use Kepler's third law and results obtained in the last question to obtain the resulting increase in time period.

---

*For a circle of course $a = b = R$ and $e = 0$.
†The area A enclosed by a closed conic section (circle or ellipse) is given by $A = \pi ab$.

3.3 Conclusion

In this chapter, we have concentrated on techniques of visualization and representation of rotating systems. Our aim has been to see how the mathematics of rotation emerges from the physics. By introducing the relationships between *rotated* coordinate systems in a matrix representation, it was possible to approach *rotating* frames of reference using the same mathematical framework, which provides a neat and tidy structure. The vector product appears in this structure quite naturally, and a generalized form for a vector description of rotation with respect to a Galilean frame is achieved. This is made possible by the introduction of the *angular velocity* vector ω, which represents the magnitude (rate) of rotation as well as the direction of rotation.

All of this is then applied to elucidate problems in 'fictitious forces' and the derivation of planetary trajectories from Newtonian mechanics. This chapter is also linked closely with workshops contained in Sections 7.8–7.11 in Chapter 7. In order to fully appreciate those workshops, students meeting the material for the first time should work through Chapter 3 first.

4
Oscillations and waves

Two of the greatest frameworks in physics have already been described in this book. Classical mechanics predicts the general motion of any body (whether in straight lines or rotation), and the study of fields enables us to understand the interaction of different objects, and as such goes to the root of what physics is about. And yet, the story, if left there, would be incomplete even for a nineteenth-century physicist. Very often (especially in practical situations) fields conspire (through the principles of mechanics) to produce repeating patterns of motion. While these can be studied using the normal techniques of classical mechanics, special methods for describing them make life much simpler for the physicist or engineer.

The physicist appreciates these methods because they often illuminate a new facet of nature. When Maxwell combined the electromagnetic field equations to produce the possibility of an electromagnetic oscillation (i.e. wave), he opened the door to the development of radio, and thus to all manner of communications we enjoy today. When Young, Fraunhofer, Fresnel, and others applied wave methods to light, they gave the study of light a level of importance which has not died to this day. When de Broglie, Schrödinger, and others accredited particles themselves to have wave properties, it gave birth to an exceptionally fruitful form of quantum theory, which now gives the theoretical underpinning to much of chemistry.

To the engineer, the wave methods hold a different fascination. Vibrations occur for a whole variety of reasons. Unless these can be understood, the safety of a structure cannot be ensured. A bridge which is perfectly stable in still air may not be so secure if subjected to periodic forces. In addition, the electronic engineer understands that without a very sound comprehension of waves, no communication system can be developed effectively.

4.1 Describing an oscillation

If an object oscillates, and we draw a graph of its position as a function of time, the result (for a 2 Hz oscillation) is shown in Figure 4.1.

The graph can be described using the sine function, since the position is zero (and increasing) at $t = 0$. However, $y = \sin t$ would not describe this oscillation, since $\sin t$ goes through one whole cycle every time t gets bigger by 2π radians, whereas our wave goes through one whole cycle every half second. Accordingly, the function we need to use is $y = \sin 4\pi t$. Similarly, a wave with a frequency of 1 Hz would be described by the function $y = \sin 2\pi t$. Usually we would write this as

$$y = A \sin \omega t, \tag{4.1}$$

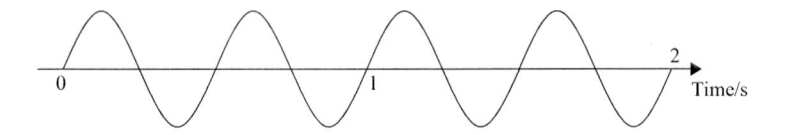

Fig. 4.1

where A is the amplitude and ω is called the angular frequency. The angular frequency is related to the frequency f and time period T by the equations:

$$\omega = 2\pi f, \tag{4.2}$$
$$\omega = 2\pi/T. \tag{4.3}$$

We have not finished with the simple oscillation yet. Not all oscillations start with $y = 0$ when $t = 0$. Some have a peak at $t = 0$, and for these we may write $y = A\cos\omega t$. However, others start neither at a peak nor at the equilibrium position. For these, we introduce a phase factor ϕ which tells the 'angle' of the cosine when $t = 0$. In general then, we may write an oscillatory motion as

$$y = A\cos(\omega t + \phi). \tag{4.4}$$

This can be shown to be equivalent to

$$\begin{aligned} y &= A\cos(\omega t + \phi) \\ &= A\cos\omega t \cos\phi - A\sin\omega t \sin\phi. \end{aligned} \tag{4.5}$$

Now if you have a free choice of A and ϕ, you can choose the 'amplitudes' of the sine and cosine terms to be any values you like. Accordingly, you will frequently see the general solution written as

$$y = C\cos\omega t + D\sin\omega t, \tag{4.6}$$

where the constants C and D are chosen to give the right overall amplitude and phase to the oscillation. The A and ϕ of equation (4.4) are related to the C and D of (4.6) by the equations:

$$\begin{aligned} C &= A\cos\phi \\ D &= -A\sin\phi \end{aligned} \tag{4.7}$$

$$\begin{aligned} A &= \sqrt{C^2 + D^2} \\ \phi &= \tan^{-1}\left(\frac{-D}{C}\right).^* \end{aligned} \tag{4.8}$$

*This formula for ϕ must be used with care, since there will be two solutions for ϕ within the range $0 \leq \phi < 2\pi$. The correct one must be chosen using common sense or by checking using equation (4.7). An example of this checking is given in workshop 4.1.1 part (a)(iv).

80 Oscillations and waves

If an object's motion is described by the equation $y = A\cos(\omega t + \phi)$, then we may work out its speed and acceleration:

$$u = \frac{dy}{dt} = -\omega A \sin(\omega t + \phi),$$
$$a = \frac{d^2 y}{dt^2} = -\omega^2 A \cos(\omega t + \phi) = -\omega^2 y. \qquad (4.9)$$

You will notice that the acceleration is proportional to the displacement y but has opposite sign. Motion such as this is called simple harmonic motion, and an oscillator undergoing this kind of motion is said to be a harmonic oscillator.

Notice that it follows from equation (4.9) that

$$u^2 + \omega^2 y^2 = \omega^2 A^2 \qquad (4.10)$$

at all times. When equation (4.10) is multiplied by half the mass of the oscillator, the left term gives the kinetic energy, the second term the potential energy, and the final term the constant total energy. This makes it clear that the energy of a wave is proportional to the square of its amplitude.

Q1 Write down an equation for an oscillation with frequency 4 Hz, where the amplitude is 4 cm, and the object starts off with $y = 4$ cm.

Q2 Write down an equation for an oscillation with frequency 10 Hz, where the amplitude is 5 cm, and the objects starts off with $y = 3$ cm (and getting bigger).

Q3 Work out the amplitude and phase factor of an oscillation described by the equation $y = 3\cos\omega t + 4\sin\omega t$.

Q4 An object moves in a two-dimensional plane where its position vector is given by $(x,y) = (5\sin\omega t, 5\sin 2\omega t)$. Describe its motion – and use a graphical calculator or spreadsheet to check your answer. +

4.1.1 Workshop: Simple harmonic motion

Motion of the form described in equation (4.9) is so important that it is worth further study. Any object which remains stationary is in equilibrium. If you slightly displace such an object with a knock or a shake, it usually has a tendency to return to its earlier position with a force more or less proportional to the disturbance. These conditions give rise to simple harmonic motion.

(a) Suppose an object with mass $m = 0.3$ kg is free to move in one dimension, with its position at any time given by the function $y(t)$. Suppose in addition that it is subject to a force which is proportional to its distance from the $y = 0$ point. If the force acting on the object has the same sign as y, the force will point away from the equilibrium point and the object will never return. However, if the force acting on the object points back towards $y = 0$, the force will have the opposite sign to y and oscillations will occur.

For these questions, assume that the force on the mass is given by $F = -ky$ where $k = 1.2$ N/m.

  (i) Show that Newton's second law for this oscillator takes the form of equation (4.9) and find the angular frequency of the motion.
  (ii) Suppose that the object is stationary at $t = 0$, with $y(0) = 2$ m. Assume that the motion can be described by an equation of the form of (4.6), and calculate the C and D coefficients. What is the amplitude of the motion?
  (iii) Repeat part (ii) assuming that the object is at $y = 0$ at $t = 0$, but with a velocity of $+6$ m/s.
  (iv) Suppose that at $t = 0$, $y = 2$ m and $dy/dt = -1.5$ m/s. Derive an equation to describe the motion at later times, and convert it to the form $y = A\cos(\omega t + \phi)$. From this write down the amplitude of the motion, and work out the first time after $t = 0$ at which the object is stationary.

(b) We now extend the work of section (a) to include the situation where the force on the mass has a part proportional to y and a constant part. We write $F = -ky + h$ where you may take $k = 1.2$ N/m and $h = 0.45$ N. In this workshop we shall satisfy ourselves that the frequency of oscillation is not affected by the presence of the constant term h.

  (i) Write Newton's second law for this oscillator and show that it takes the form:
  $$\frac{d^2 y}{dt^2} = Py + Q,$$
  where P and Q are constants, which you should evaluate in terms of m, k and h.
  (ii) Show that $y = A\cos(\omega t + \phi)$ can never be the solution to your answer to (i) no matter what values are chosen for A, ω, or ϕ.
  (iii) Derive an expression for the position of the equilibrium point y_0, that is the point at which the mass experiences zero resultant force.
  (iv) If we define new coordinates y' which measure the position of the object with respect to the equilibrium position (so that $y' = y - y_0$), show that Newton's second law in the new coordinates has no constant Q term, and accordingly, y' can be expressed in the form $A\cos(\omega t + \phi)$.

(c) Suppose that our 0.3 kg mass is subject to a force $F = -ky + hy^2$ where $k = 1.2$ N/m as before and $h = 2$ N/m^2. The mass is initially at rest at $y = 0$. It is then knocked out of position by 3 cm and released. Will it undergo simple harmonic motion?

4.2 Workshop: Introducing complex numbers

It turns out that there is a simpler way of representing oscillation than using sines and cosines – and this involves the use of complex numbers. Accordingly in this section we revise the basic ideas of complex numbers.

We start with a number line, as shown in Figure 4.2, with the zero in the middle and the positive numbers to its right. We now imagine this to be the x-axis of a

82 Oscillations and waves

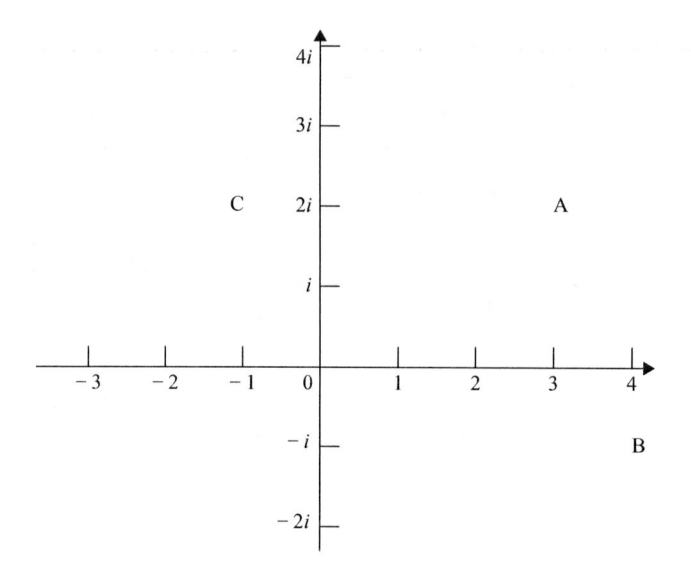

Fig. 4.2

two dimensional surface, and it is called the real axis. The corresponding y-axis is called the imaginary axis. The number one unit along the positive y-axis from zero is called i, the next is twice as big ($2i$), then we have $3i$, $4i$, and so on.* Similarly, on the negative y-axis we have $-i$, $-2i$, $-3i$, and so on.

Any point on the plane can be described by its x- and y-coordinates, which are called the real and imaginary parts of the number. Thus the point A in Figure 4.2 is $3 + 2i$, while B is at $4 - i$ and C is at $-1 + 2i$.

Complex numbers can be added or subtracted in a manner analogous to the addition or subtraction of vectors, where real and imaginary parts of numbers are summed separately. Thus $(3 + 2i) + (-1 + 6i) = 2 + 8i$.

(a) What happens to a point on the complex plane when you add 1 to it?
(b) What happens to a point on the complex plane when you add i to it?
(c) Describe what happens to a point on the complex plane when you multiply it by 2 (this multiplies real and imaginary parts by 2).
(d) Describe what happens to the point labelled $+1$ when you multiply its complex number by i. What happens when you multiply the point labelled -1 by i?

You should have found that multiplication of a complex number by a real number (like 2) moves it away from or towards zero, without changing its 'bearing' from the origin. On the other hand, multiplication by i takes the point representing a real number and rotates it by 90° anticlockwise with the origin as the centre of rotation.

Warning to engineers: Frequently you will see the letter j used instead of i. This is because in engineering i is often used to label electric currents, and we do not want to get them confused. That said, things are rarely this straightforward. The letter j is frequently used to label currents too.

If we want multiplication by i to rotate *any* point by 90° anticlockwise, then this means that we must require that $i \times i = -1$.

(e) Start with a point $z = x + iy$, where x and y are real (they label the real and imaginary parts of z). Using the information that $i \times i = -1$, show that $iz = -y + ix$, and accordingly that this point has also been rotated by 90° about the origin on multiplication by i. Note that we sometimes use the notation $\operatorname{Re}(z)$ to mean 'real part of z', and similarly $\operatorname{Im}(z)$ means 'imaginary part of z'.

(f) Show that, for any complex numbers a and b, $\operatorname{Re}(ab) = \operatorname{Re}(a)\operatorname{Re}(b) - \operatorname{Im}(a)\operatorname{Im}(b)$ and that $\operatorname{Im}(ab) = \operatorname{Re}(a)\operatorname{Im}(b) + \operatorname{Im}(a)\operatorname{Re}(b)$. *Hint*: Write a and b in terms of their real and imaginary parts ($a = p + iq$, $b = r + is$), and then work out ab in terms of p, q, r, and s.

(g) The complex conjugate of a number is defined to be the reflection of its point in the real axis. Thus the complex conjugate of $z = x + iy$ is $z^* = x - iy$ (here, once more, we take x and y to be real numbers). Write down the complex conjugates of $3 + 2i$, $5 - 3i$, $-2 + i$, and $-4 - 5i$.

(h) Show that if $z = ab$, where z, a, and b are all complex, then $z^* = a^*b^*$.

(i) Show that $\operatorname{Re}(z) = \frac{1}{2}(z + z^*)$, and find a similar expression for $\operatorname{Im}(z)$.

(j) The 'direct distance' from the origin to a point on the complex plane is called its modulus. Thus the modulus of $3 + 4i$, written as $|3 + 4i| = 5$. Show that when you multiply a complex number z by its complex conjugate z^*, you always get a real number equal to $|z|^2$. For this reason z^*z is called the modulus square of z.

(k) The 'bearing' of a point on the complex plane is called its argument (Arg). By convention, the arguments of positive, real numbers are zero; the arguments of positive imaginary numbers are $+\pi/2$ (never, but *never* use degrees). If $\operatorname{Arg}(z) = \theta$, show that $\operatorname{Re}(z) = |z|\cos\theta$, while $\operatorname{Im}(z) = |z|\sin\theta$.

(l) Work out the real and imaginary parts of i^3, i^4, i^5, and i^6.

(m) It can be shown (see Section 7.2) that

$$e^x = 1 + x + \frac{x^2}{2!} + \frac{x^3}{3!} + \cdots.$$

It is also true (providing your angles are in radians, as they should be) that

$$\cos\theta = 1 - \frac{\theta^2}{2!} + \frac{\theta^4}{4!} - \frac{\theta^6}{6!} \cdots \quad \text{and} \quad \sin\theta = \theta - \frac{\theta^3}{3!} + \frac{\theta^5}{5!} - \frac{\theta^7}{7!} \cdots.$$

Use these facts, together with your answers to part (l) to show that $e^{i\theta} = \cos\theta + i\sin\theta$. This means that $re^{i\theta}$ is a natural way to represent the complex number which has modulus r and argument θ.

(n) If I multiply the complex number $re^{i\theta}$ by $se^{i\phi}$, what is the result? Assume that r, s, ϕ, and θ are all real. What is its modulus? What is its argument?

(o) Suppose $z = e^{i\theta}$ and $w = e^{i\phi}$, where θ and ϕ are real. Show that $\operatorname{Re}(zw) = \cos(\theta + \phi)$ using your results to parts (f), (k), and (n). Remembering that $\operatorname{Re}(z) = \cos\theta$, $\operatorname{Re}(w) = \cos\phi$, and so on, show that $\cos(\theta + \phi) = \cos\theta\cos\phi - \sin\theta\sin\phi$.

84 Oscillations and waves

Similarly, use the imaginary parts to prove that $\sin(\theta + \phi) = \sin\theta\cos\phi + \cos\theta\sin\phi$. *Hint*: You may find your answers to (f) or (i) useful.

(p) Show that $\cos\theta = \frac{1}{2}\left(e^{i\theta} + e^{-i\theta}\right)$ and that $\sin\theta = \frac{1}{2i}\left(e^{i\theta} - e^{-i\theta}\right)$.

4.3 Describing an oscillation using complex numbers

If we use the results of the last section, we can rewrite our equation (4.4) in terms of the real part of a complex number:

$$y = \operatorname{Re} A e^{i(\omega t + \phi)}. \tag{4.11}$$

This in turn can be written as

$$y = \operatorname{Re} A e^{i\phi} e^{i\omega t}, \tag{4.12}$$

and if we define a complex amplitude $B = A e^{i\phi}$, then our equation becomes

$$y = \operatorname{Re} B e^{i\omega t}. \tag{4.13}$$

Complex numbers are ideally suited to describing oscillations, because a complex number, with its modulus and argument, can describe both amplitude and phase at the same time.

The convenience of using complex numbers to describe oscillations becomes clear if we wish to add two oscillations together. Suppose that $y = A\cos\omega t + B\cos(\omega t + \phi)$. We can simplify this by looking up trigonometric identities; however, it is much simpler to note that

$$y = \operatorname{Re} A e^{i\omega t} + \operatorname{Re} B e^{i(\omega t + \phi)}$$
$$= \operatorname{Re}\left\{\left(A + B e^{i\phi}\right) e^{i\omega t}\right\}, \tag{4.14}$$

and that the complex amplitude of the combined wave is given by the number $A + Be^{i\phi}$. The amplitude and phase of the oscillation are then given by the modulus and argument of this complex amplitude. They can be worked out as

Amplitude:
$$\left|A + Be^{i\phi}\right| = \sqrt{(A + Be^{i\phi})(A + Be^{-i\phi})}$$
$$= \sqrt{A^2 + B^2 + AB(e^{i\phi} + e^{-i\phi})}$$
$$= \sqrt{A^2 + B^2 + 2AB\cos\phi}. \tag{4.15}$$

Phase:
$$\operatorname{Arg}\left(A + Be^{i\phi}\right) = \tan^{-1}\left(\frac{\operatorname{Im}(A + Be^{i\phi})}{\operatorname{Re}(A + Be^{i\phi})}\right)$$
$$= \tan^{-1}\left(\frac{B\sin\phi}{A + B\cos\phi}\right) \tag{4.16}$$

While this may seem a trifle messy, it is not as messy as using the trigonometric functions throughout. In addition, we are usually happy to calculate the complex amplitude and leave it at that – it does after all contain all of the essential information.

Notice one other helpful consequence of using complex numbers to describe an oscillation. Suppose that $y = A \cos \omega t = \operatorname{Re} A e^{i\omega t}$. It follows that

$$\frac{dy}{dt} = \operatorname{Re} i\omega A e^{i\omega t}, \tag{4.17}$$

and hence that the complex amplitude of the *velocity* is directed along the imaginary axis of the complex plane at $t = 0$, showing that it has been rotated by $\pi/2$ radians with respect to the displacement y. This in turn means that the derivative has a phase $\pi/2$ greater than the original, and we can observe this in much greater clarity than if we had differentiated $A \cos \omega t$ to get $-A\omega \sin \omega t$. Waves of different phases can be added very conveniently without the need to add sines to cosines – and this is a great strength of the notation, as will become clear when we look at interference calculations.

4.4 Workshop: Damped oscillators

To demonstrate the usefulness of complex numbers to describe oscillations in practical situations, we are now going to solve the 'damped oscillator' problem. This describes the motion of a car wheel on the end of a suspension point, a bridge when a load is applied, and many other things besides.

Suppose a particle has mass m and is constrained to move in one dimension only. Its position is given by x, and it is subject to two forces:

- A restoring force directed towards $x = 0$ of magnitude kx.
- A friction force directed in opposition to its motion, which is proportional to its speed. The force has magnitude $r\dot{x}$, where we use the dot to signify 'time derivative'.

(a) Show that the equation satisfied by the motion of the particle is

$$F = ma$$
$$-kx - r\dot{x} = m\ddot{x},$$

and hence that $m\ddot{x} + r\dot{x} + kx = 0$.

(b) Now we assume that the solution is given by $x = \operatorname{Re}(z)$, where z is a complex number equal to $Ae^{\alpha t}$ and A and α are complex constants as yet unknown. Show that our solution will suffice providing that

$$\left(m\alpha^2 + r\alpha + k\right) z = 0. \tag{4.18}$$

(c) Given that we do not want $z = 0$, it follows that we require a value of α such that the quadratic in brackets is zero. Find the two values of α which satisfy this condition – we shall call them α_1 and α_2.

If $r^2 - 4mk > 0$, then there is a lot of resistance in the system – it is said to be overdamped. In this case α_1 and α_2 are both real and negative, and the solution can be written as

$$z = A_1 e^{\alpha_1 t} + A_2 e^{\alpha_2 t}. \tag{4.19}$$

86 Oscillations and waves

If $r^2 = 4mk$, then the situation is said to be critically damped, and the solution is of the form

$$z = e^{-rt/2m} (A_1 + A_2 t). \tag{4.20}$$

(d) Show that the solution given as equation (4.20) does satisfy the differential equation in the case when $r^2 = 4mk$.

(e) If $r^2 - 4mk < 0$, we have an underdamped situation. Remembering that $i^2 = -1$, show that α_1 and α_2 are now equal to

$$-\frac{r}{2m} + i\sqrt{\frac{k}{m} - \frac{r^2}{4m^2}} \quad \text{and} \quad -\frac{r}{2m} - i\sqrt{\frac{k}{m} - \frac{r^2}{4m^2}}. \tag{4.21}$$

(f) In this case, show that the solution is of the form

$$z = e^{-rt/2m} \left(A e^{i\omega t} + B e^{-i\omega t} \right), \tag{4.22}$$

where ω is equal to the square root in (4.21), and is real. The solution can also be written in the form

$$x = e^{-rt/2m} \left(C \cos \omega t + D \sin \omega t \right), \tag{4.23}$$

where C and D are real. The solution is an oscillation with diminishing amplitude.

(g) Try solving parts (e) to (f) without using complex numbers. Assume that you already know that the solution is of the form $e^{-at} (C \cos \omega t + D \sin \omega t)$, and determine the values of a and ω in terms of m, k, and r.

4.5 Describing a wave in one dimension

If you set up a row of oscillators and tie them together, then each one causes the next one to move. This sets up a wave. Let us suppose that a wave is being set up by an oscillator at a position we call $x = 0$, which moves according to $y = A \cos(\omega t + \phi)$ (Figure 4.3). The peaks and troughs then move away from this oscillator to increasing x at a speed c.

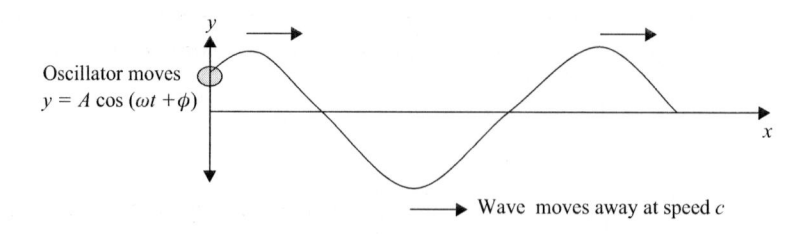

Fig. 4.3

I want to know the height y of the wave at position x at time t. How do I work this out?

Whatever I have now at position x has taken a time x/c to reach this point from the oscillator. Accordingly whatever is at position x now was at the origin x/c seconds ago. Thus whatever is at position x at time t was at the origin at time $t - x/c$. Mathematically, this reasoning is written as

$$y(x,t) = y\left(0, t - \frac{x}{c}\right) = A\cos\left\{\omega\left(t - \frac{x}{c}\right) + \phi\right\} = A\cos\left(\omega t - \frac{\omega x}{c} + \phi\right). \quad (4.24)$$

We now look at this function for a fixed value of t. This is equivalent to taking a 'freeze frame' picture of the wave in progress. Let us find its wavelength. As you look further down the wave, the wave will start repeating its pattern once $\omega x/c$ has increased by 2π. To increase $\omega x/c$ by 2π, x must be increased by $2\pi c/\omega$. Therefore the wavelength $\lambda = 2\pi c/\omega$. However as $\omega = 2\pi f$, this is equivalent to the more familiar equation $\lambda = c/f$. We can rewrite our wave using the wavelength as

$$y = A\cos\left(\omega t - \frac{2\pi x}{\lambda} + \phi\right), \quad (4.25)$$

although it is more common to define a wavenumber $k = 2\pi/\lambda$, and then to write

$$y = A\cos(\omega t - kx + \phi) \quad (4.26)$$

or

$$y = \operatorname{Re} A e^{i\phi} e^{i(\omega t - kx)}. \quad (4.27)$$

The use of k is partly done to make the equation more tidy – however, it has a knock on benefit that the quantity k can be considered a vector when the wave is moving in three dimensions. We shall come back to this later.

Before leaving this section of wave description, it is worth making one point. If you change the $-kx$ into a $+kx$, you make a wave which moves the other way: towards $-x$.

Q5 Show that ω/k gives the wave speed c.
Q6 Work out the wavenumber and angular frequency for a 2 m wavelength, 7 Hz wave. If the wave has an amplitude of 3 cm, write an equation to describe it. Assume that there is a maximum (peak) at $x = 0$ when $t = 0$.
Q7 Write the equation for a wave identical to that in Q6, only going the opposite way.
Q8 Add the two waves you wrote in Q6 and Q7. This makes a standing wave. There are certain values of x for which $y = 0$ at all times. These are called the nodes. Calculate the position of the nodes. Derive an expression for the amplitude of the oscillations at position x. +

4.6 Interference – a brief introduction

Most wave physics involves encouraging a wave to take more than one route from a source to a detector, and then experimenting and calculating to find the intensity

88 Oscillations and waves

of the wave at the detector. The different routes will have different lengths. If the original wave started off as $y = \text{Re}\left(Be^{i(\omega t - kx)}\right)$, then after travelling along an additional path of length L, it will now be described by $y = \text{Re}\left(Be^{i(\omega t - k(x+L))}\right) = \text{Re}\left(Be^{-ikL}e^{i(\omega t - kx)}\right)$.

Let us now suppose that two waves of equal amplitude B arrive at the same point (same x) via two different routes of lengths L_1 and L_2. The total disturbance is equal to the sum of the contributing wave displacements:

$$y = \text{Re}\left(Be^{-ikL_1}e^{i(\omega t - kx)} + Be^{-ikL_2}e^{i(\omega t - kx)}\right)$$
$$= \text{Re}\left(B\left(e^{-ikL_1} + e^{-ikL_2}\right)e^{i(\omega t - kx)}\right). \qquad (4.28)$$

The complex amplitude is thus given by $B\left(e^{-ikL_1} + e^{-ikL_2}\right)$. We can calculate the measured amplitude if we first calculate the modulus square of the complex amplitude:

$$B\left(e^{-ikL_1} + e^{-ikL_2}\right) \times B^*\left(e^{ikL_1} + e^{ikL_2}\right)$$
$$= |B|^2\left(2 + e^{ik(L_1 - L_2)} + e^{ik(-L_1 + L_2)}\right)$$
$$= |B|^2\left(2 + 2\cos k(L_2 - L_1)\right)$$
$$= |B|^2 4\cos^2\left(\frac{1}{2}k(L_2 - L_1)\right),$$

which gives us a real amplitude (modulus) of

$$2\left|B\cos\left(\frac{1}{2}k(L_2 - L_1)\right)\right|. \qquad (4.29)$$

This depends on the difference in the two path lengths. You will often see $L_2 - L_1$ written as the 'path difference' or 'path length difference'. Once multiplied by k, the path difference becomes the phase difference of the two waves. We see from equation (4.29) that if the phase difference is an even multiple of π the resulting amplitude will be twice the original amplitude. If the phase difference is an odd multiple of π, there will be complete destructive interference, and the waves will 'cancel out'.

Q9 Suppose three waves of equal amplitude meet at a point. The complex amplitudes of the three waves are $Ae^{-i\phi}$, A, and $Ae^{i\phi}$, respectively. Calculate the modulus square amplitude, and find which values of ϕ give rise to full constructive interference and full destructive interference. +

Q10 Use the formula for the sum of a geometric series to find the modulus square amplitude when N waves meet at a point, where the complex amplitude of the pth wave is given by $Ae^{ip\phi}$, where p can take the values $0, 1, 2, 3, \ldots, N-1$. Determine the values of ϕ for which full constructive and full destructive interference occur. ++

Q11 Plot the modulus square amplitudes as functions of ϕ for Q9 and Q10 in the cases where $N = 3$, $N = 5$, and $N = 10$. What would you expect to find for $N = 10\,000$?

4.7 Workshop: The wave equation

So far we have looked at how you describe a wave, but we have said nothing about how they 'work'. It turns out that whenever you have a wave, the physics of the situation (be it water waves on the ocean, seismic waves in the Earth or radio waves in the air) fits a differential equation called the wave equation. Our equation (4.26) is the solution to the wave equation.

The wave equation is an example of a partial differential equation – that is an equation containing partial derivatives. So, in this section we shall not only introduce a vital equation but also revise our understanding of partial derivatives from Section 2.3.

We shall start with a wave of the form $y = A\cos(\omega t - kx + \phi) = \operatorname{Re} Ae^{i\phi} e^{i(\omega t - kx)}$. In the questions that follow, it is probably best if you attempt them twice – once using trigonometric functions and once using the complex exponential method.

(a) Work out the partial derivative $\partial y/\partial x$. This means that you differentiate y with respect to x, treating t as a simple constant.
(b) Differentiate your answer to (a) to get the second derivative of y with respect to x, that is $\partial^2 y/\partial x^2$. Show that this is equal to $-k^2 y$.
(c) Work out the partial derivative $\partial y/\partial t$. Here you differentiate y with respect to t, treating x as a constant.
(d) Work out the second derivative $\partial^2 y/\partial t^2$ by differentiating your answer to (c) by t once more. Show that $\partial^2 y/\partial t^2 = -\omega^2 y$.
(e) Show that $(\partial^2 y/\partial x^2) = (k^2/\omega^2)(\partial^2 y/\partial t^2) = (1/c^2)(\partial^2 y/\partial t^2)$, where $c = \omega/k$ is the speed of the wave. This partial differential equation is called the wave equation.
(f) Show that any function of the form $f(\omega t \pm kx)$ would satisfy the wave equation given in (e). While the cosine or complex exponential function is not the only solution to the wave equation, it turns out to be the most useful because it has a well-defined frequency and wavelength, and furthermore any other function can be built up from sine and cosine waves.

4.8 A wave on a string

The simplest kind of wave to analyse is a transverse wave set up on a taut string. We start with the string lined up horizontally with the x-axis, and then wiggle one end up and down in the y-direction. We shall assume that the tension in the string is T, and that this is sufficiently high that additional tensions caused by the extra stretching of the string as it wiggles are negligible in comparison to T. The position of the string at position x at time t is denoted $y(x,t)$. While the true string is continuous, it is easier to analyse if we start by thinking of it as a string of beads, each of mass m, on a light string. In this way, the string joining each bead will be straight. Let the angle between the horizontal and a string leaving a bead be α_x.

As shown in Figure 4.4, the vertical component of the force on oscillator x is given by

$$F_y = T\sin\alpha_x - T\sin\alpha_{x-\delta x},$$

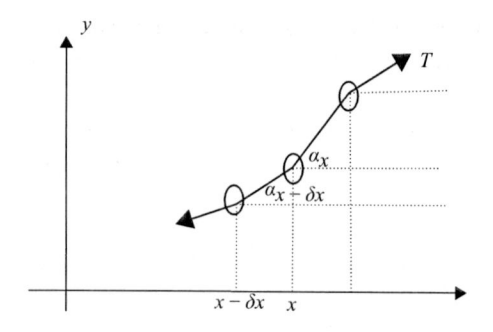

Fig. 4.4

while the horizontal component is

$$F_x = T\cos\alpha_x - T\cos\alpha_{x-\delta x}.$$

It turns out to be quite acceptable to assume that the angles α are small, and so $\cos\alpha = 1$ (to first order in α), while $\sin\alpha = \alpha$. Accordingly, all of the interesting motion will be in the vertical direction.

$$F_y = T(\alpha_x - \alpha_{x-\delta x}) \approx T\frac{\partial \alpha}{\partial x}\delta x. \quad (4.30)$$

Now, since α is small, we may also write $\alpha \approx \tan\alpha = \partial y/\partial x$. Thus,

$$F_y \approx T\frac{\partial}{\partial x}\left(\frac{\partial y}{\partial x}\right)\delta x = \frac{\partial^2 y}{\partial x^2}\delta x. \quad (4.31)$$

The bead's vertical component of acceleration is $\partial^2 y/\partial t^2$, and so we write Newton's second law for this oscillating particle as

$$F_y = ma_y$$
$$T\frac{\partial^2 y}{\partial x^2}\delta x = m\frac{\partial^2 y}{\partial t^2} \quad (4.32)$$
$$T\frac{\partial^2 y}{\partial x^2} = \frac{m}{\delta x}\frac{\partial^2 y}{\partial t^2}.$$

Now $m/\delta x$ is the mass per unit length of string, usually called the density (even though it is not a mass per unit volume) and denoted ρ. The equation then becomes

$$\frac{\partial^2 y}{\partial x^2} = \frac{\rho}{T}\frac{\partial^2 y}{\partial t^2}, \quad (4.33)$$

which is identical in form to the wave equation derived in the workshop, providing that the c^2 of the wave equation is equal to T/ρ. This not only tells us that waves can pass up and down the string, but also tells us that their speed must be given by $c = \sqrt{T/\rho}$.

Q12 Use the equation $c = \sqrt{T/\rho}$ for a transverse wave on a string to work out the mass per unit length needed on each of the strings on a violin. Violin strings are about 50 cm long, and held at a tension of about 50 N. The four strings on the instrument produce sounds of frequency 196, 293, 440, and 660 Hz, respectively. Why do you think that violin makers prefer all four strings to be at approximately equal tensions?

4.9 Energy content of a wave

Waves may oscillate, and may travel up and down a string. But at the end of the day, we use waves to carry energy, and hence information. Ascertaining the rate at which a wave carries energy is essential. The simplest way of doing this is to work out how much energy each metre's worth of wave carries. Because the wave's speed (c) is the distance covered by the signal each second, the energy passing a point in *one second* is equal to the energy contained in c metres.

If we take the wave's function as $y = A\cos(\omega t - kx)$, then the vertical component of velocity is given by $\partial y/\partial t = -A\omega \sin(\omega t - kx)$. The kinetic energy of a short length δx of string (of mass $\rho\, \delta x$) is accordingly:

$$\delta K = \frac{1}{2}\rho\, \delta x\, (\partial y/\partial t)^2 = \frac{1}{2}\rho A^2 \omega^2 \sin^2(\omega t - kx)\, \delta x. \tag{4.34}$$

The average value of the $\sin^2$ function is one half, and accordingly we expect the kinetic energy carried by a short section of wave to be (on average)

$$\delta K = \frac{1}{4}\rho A^2 \omega^2 \delta x. \tag{4.35}$$

The potential energy carried by each metre of wave is more tricky to calculate. However, if we think of the wave as a string of oscillators, any one of which is transferring its kinetic energy into potential and then back again, it follows that the average value of potential energy will be the same as the average kinetic energy.* Accordingly, the total energy of each metre is twice the value in equation (4.35) once the potential energy is included as well.

We can now calculate the power of the wave – the energy passing a point each second – by multiplying this figure by the speed of the wave:

$$P = \frac{1}{2}\rho A^2 \omega^2 \times c. \tag{4.36}$$

This equation gives us two important pieces of information. First, the power transmitted by a wave is proportional to the square of its amplitude. This means that if you want to double the power of a wave you only need to increase its amplitude by $\sqrt{2}$.

---

*We are here hiding the full details. A travelling wave is not like a string of independent oscillators in that energy can be passed from one to another down the string. It does turn out to be true that the average potential energy is equal to the average kinetic energy; however, the places of greatest kinetic energy are also the places of greatest potential energy, while the points in between have neither form of energy.

92 Oscillations and waves

Second, the equation contains the factor ρc. This proves to be fundamental in its own right, and it is called the impedance of the wave, and is given the symbol Z. While we use this information below, a fuller description of the meaning of impedance will have to wait until our chapter on electric circuits. In terms of impedance,

$$P = \frac{1}{2} Z A^2 \omega^2. \qquad (4.37)$$

Q13 Show that the impedance of a wave on a string is also given by the expression T/c.
Q14 Show that the power of a wave is also given by the expression $\frac{1}{2} T c k^2 A^2$.
Q15 When a wave moves from one medium to another, the frequency remains the same, but the wavelength usually changes. Suppose two strings are tied end to end, and a wave passes across the join from one to the other. The tensions in the strings must be equal. Show that the ratio of the impedances of the waves on the two strings is the same as the ratio of the wavenumbers k. +

While we have defined $Z = \rho c$, the impedance has another important meaning. Equation (4.37) gives the average rate of transfer of energy down the string. The actual rate fluctuates as a $\sin^2$ function with equation (4.37) as its average value. With this being the case, the instantaneous power is given by $Z\left(A\omega \sin \omega t\right)^2$. Now, as shown in equation (2.9), $P = \mathbf{F} \cdot \mathbf{v}$. Since $A\omega \sin \omega t$ is the velocity of the wave in the y-direction, it follows that $ZA\omega \sin \omega t$ must be the transverse component of the force transmitted down the string. And so the force is equal to the impedance Z multiplied by the velocity of the waving string. This fact has great significance which will be used in our next chapter when we discuss electric circuits.

4.10 Impedance matching

Engineers and other practical folk often need to make the transition of wave from one material to another as efficient as possible. To give examples, a violin maker wants to get as much of the vibration of the string into the air (as music) as possible. An electric guitarist at a gig might use a direct input (DI) box to improve the efficiency of the transfer of electrical 'sound' from the guitar to the amplification system. There are different sockets on the back of a TV depending on which kind of aerial you have. Before a medic does an ultrasound scan in a hospital, they put a gel between the probe and the skin. There is no point connecting 200 Ω speakers to an amplifier which is designed for 8 Ω speakers – you will not hear much. And if you are using an oscilloscope to watch for a finely calibrated pulse in the laboratory, it is handy to set the input impedance to 50 Ω rather than 'high' otherwise you get reflected pulses which confuse your experimental equipment.

The source of the problem is that when a wave moves from one material to another, some will reflect at the boundary, and this is a waste of energy or signal. By studying the behaviour of waves as they cross boundaries between materials of different impedance, we can learn how to reduce the reflection.

The simplest kind of transfer is where we tie strings of different thickness together, and then make a wave on the string. When the wave hits the join, we can work out how

Impedance matching 4.10

much will reflect and how much will go over to the other string. For convenience, let us say that the join is at $x = 0$, and the waves are coming from a source at negative x. The impedance of the material is Z_L for $x < 0$ and Z_R for $x > 0$. We shall ignore the extra mass of the knot, and write separate equations for the waves involved.

Original (incident) wave:

$$y_i = A_i \cos(\omega t - k_L x) \quad \text{for } x < 0 \text{ only.} \tag{4.38}$$

Reflected wave:

$$y_r = A_r \cos(\omega t + k_L x) \quad \text{for } x < 0 \text{ only.} \tag{4.39}$$

The $+$ shows that this wave is moving to the left. The actual shape of the wave for $x < 0$ will be given by $y_i + y_r$ using the principle of wave superposition. So the full equation for $x < 0$ is

$$y_i + y_r = A_i \cos(\omega t - k_L x) + A_r \cos(\omega t + k_L x) \quad \text{for } x < 0. \tag{4.40}$$

For $x > 0$, the only wave is the transmitted wave:

$$y_t = A_t \cos(\omega t - k_R x) \quad \text{for } x > 0 \text{ only.} \tag{4.41}$$

Note that because the two materials have different impedances, the waves will have different wavenumbers k_L and k_R depending on which material they are in.

There are two rules to use as we try and link the equations describing the left and right halves of the problem.

First, because the strings are joined at $x = 0$, the equations for $x < 0$ and $x > 0$ must agree on the value of y at the join. This means that $y_i(0,t) + y_r(0,t) = y_t(0,t)$ for all times. Accordingly, $A_i \cos \omega t + A_r \cos \omega t = A_t \cos \omega t$, and so not only must all the frequencies be the same on either side of the join (as our notation has assumed), but also

$$A_i + A_r = A_t. \tag{4.42}$$

Second, there must be no unbalanced force at the join itself, since the join has no mass, and so any unbalanced force would cause an infinite acceleration. The vertical component of force on the right of the join is $+T \sin \alpha \approx T \partial y_t / \partial x$, with T being the horizontal component of tension in the string and α being the angle made by the string to the horizontal. On the left of the join the force is similarly given by $-T \partial (y_i + y_r) / \partial x$. These two vertical components must sum to zero, giving us the following equation:

$$\frac{\partial y_i}{\partial x}\bigg|_{x=0} + \frac{\partial y_r}{\partial x}\bigg|_{x=0} = \frac{\partial y_t}{\partial x}\bigg|_{x=0}, \tag{4.43}$$

and so

$$-A_{\mathrm{i}}k_{\mathrm{L}} \sin \omega t + A_{\mathrm{r}}k_{\mathrm{L}} \sin \omega t = -A_{\mathrm{t}}k_{\mathrm{R}} \sin \omega t$$
$$(A_{\mathrm{r}} - A_{\mathrm{i}})\,k_{\mathrm{L}} = -A_{\mathrm{t}}k_{\mathrm{R}}. \tag{4.44}$$

From equations (4.42) and (4.44), we can work out

$$A_{\mathrm{r}} = A_{\mathrm{i}} \frac{k_{\mathrm{L}} - k_{\mathrm{R}}}{k_{\mathrm{L}} + k_{\mathrm{R}}}, \tag{4.45}$$

$$A_{\mathrm{t}} = A_{\mathrm{i}} \frac{2k_{\mathrm{L}}}{k_{\mathrm{L}} + k_{\mathrm{R}}}, \tag{4.46}$$

and so we get no wasteful reflection when $k_{\mathrm{L}} = k_{\mathrm{R}}$. Now because the tension and angular frequency of the wave must be the same on left and right, and because $Z = T/c = Tk/\omega$, then for the strings Z is proportional to k. Hence equations (4.45–4.46) can be written as

$$A_{\mathrm{r}} = A_{\mathrm{i}} \frac{Z_{\mathrm{L}} - Z_{\mathrm{R}}}{Z_{\mathrm{L}} + Z_{\mathrm{R}}}, \tag{4.47}$$

$$A_{\mathrm{t}} = A_{\mathrm{i}} \frac{2Z_{\mathrm{L}}}{Z_{\mathrm{L}} + Z_{\mathrm{R}}}. \tag{4.48}$$

The business of keeping k the same on left and right to ensure total transmission is the same as keeping the impedance the same. Therefore you will hear people talk about 'impedance matching' when they want to ensure efficient transmission.

Q16 Show that the efficiency of the join in terms of power (rather than amplitude) is given by $\eta = 4Z_{\mathrm{L}}Z_{\mathrm{R}}/(Z_{\mathrm{L}} + Z_{\mathrm{R}})^2$ using equations (4.37) and (4.48). Work out the fraction of energy reflected, and show that all of the incident energy is either reflected or transmitted.

4.11 Describing waves in three dimensions

All of the waves we have looked at so far have been in one dimension. Our reality is three dimensional, and so we next look at how you describe a wave in three dimensions, and we shall look at two types of three dimensional wave – the plane wave and the spherical wave.

4.11.1 Plane waves

Suppose a quantity called E is waving, with the waves moving in the $+x$-direction. We can write this wave $E = E_0 \cos(\omega t - kx)$ even though it is in three dimensions, since the oscillation is doing the same thing in step at all y- and z-values. This is a plane wave. Similarly a plane wave moving to $+y$ would be described using $E = E_0 \cos(\omega t - ky)$, and one moving to $-y$ would be $E = E_0 \cos(\omega t + ky)$.

What about a wave moving in some other direction? Let us set up a vector $\mathbf{s} = (u, v, w)$ which points in the direction that the wave is moving, and to make our mathematics easier, let us normalize $\mathbf{s}$ so that it has unit length (i.e. $u^2 + v^2 + w^2 = 1$). Let us also set up an axis through the origin which points in the s-direction.

The wave is described by $E = E_0 \cos(\omega t - kd)$, where d is the distance of the point from the origin as measured parallel to our **s**-axis. But how do we work out the value of d at a particular point $\mathbf{r} = (x, y, z)$? The answer is that d equals the component of **r** parallel to **s**, and since **s** is normalized, it follows that $d = \mathbf{r} \cdot \mathbf{s}$. We can then write a plane wave travelling in the **s**-direction as

$$E = E_0 \cos(\omega t - k\mathbf{s} \cdot \mathbf{r}). \tag{4.49}$$

It is usually more convenient to define a vector $\mathbf{k} = k\mathbf{s}$, that is a wave vector which points in the direction that the wave is travelling, and has a magnitude given by $2\pi/\lambda$. Once this is done, the plane wave can be written:

$$E = E_0 \cos(\omega t - \mathbf{k} \cdot \mathbf{r}). \tag{4.50}$$

Q17 Using the plane wave equation (4.50), show that $\partial E/\partial x = E_0 k_x \sin(\omega t - \mathbf{k} \cdot \mathbf{r})$ where k_x is the x-component of **k**.

Often waves are vectors, and so E of equation (4.50) would be a vector **E** rather than a scalar E. The electric field of a radio wave or the displacement of a molecule from its mean position during a seismic wave would be examples. In this case, we have two main classes of wave – transverse (where **E** is perpendicular to **k**) and longitudinal (where **E** is parallel to **k**). When dealing with vector waves such as **E**, it is worth remembering that there is no special link between the x-component of **E** and the x-component of **k**. E_x gives the amplitude observed in the x-direction. k_x gives the wavenumber when the wave is looked at along the x-axis alone.

4.11.2 Spherical waves

The final type of wave we shall describe is the wave which spreads out radially. If we want to know what is happening at a point (x, y, z), the important thing is its distance from the centre where the waves were made, and we call this distance r. Initially we might want to write $E = E_0 \cos(\omega t - kr)$, however, there is a problem.

Remember from equation (4.37) that the power of a wave is related to the square of its amplitude. If the energy in a ripple remains the same as the ripple spreads, the energy in each part of the ripple must get smaller (the wave spreading out). As each spherical wavefront of the ripple spreads to radius r, the area it is covering is now $4\pi r^2$, so the fraction of the wave's energy meeting each square metre at this radius is now $1/4\pi r^2$. Accordingly the amplitude must reduce by the square root of this factor, and we usually write

$$E = \frac{E_0}{r} \cos(\omega t - kr), \tag{4.51}$$

where E_0 is the amplitude of the wave 1 m from the source.

So at radius r, the intensity (or local power of the wave in W/m^2) is proportional to $(E_0/r)^2$, and the total energy passing this radius is proportional to $(E_0/r)^2 \times 4\pi r^2 = E_0^2 \times 4\pi$ which is the same as for any other radius. Accordingly, the total energy remains the same – it is just spread over a larger area. To draw on a two dimensional example, this is why ripples in a pond get less high as the radius of the ripple increases.*

Q18 The power output of the Sun is about 3.9×10^{26} W. The Sun has a radius of about 6.96×10^8 m, and the Earth is approximately 1.50×10^{11} m from the centre of the Sun. Work out (1) the intensity (in W/m²) of sunlight at a point on Earth facing the Sun directly and (2) the intensity (in W/m²) of sunlight at a point just above the Sun's own surface. Assume that none of the sunlight is absorbed in space.

4.11.3 Workshop: Stellar magnitudes

Astronomers use 'magnitudes' to refer to the brightness of stars. In ancient times, the brightest stars were put in the 'first magnitude', the next brightest in the 'second magnitude' until the 'fifth magnitude' contained those just visible to the naked eye. In subsequent years, sixth and higher magnitudes were used to label stars visible only with telescopes, while one or two stars are sufficiently bright to warrant promotion from the first to a zeroth magnitude (the 'premier league' of stars).

The brightness of a star partly depends on how far away it is. Accordingly we use absolute magnitudes (denoted M) to label how bright the star actually is, while apparent magnitudes (m) label the brightness of the star as viewed from Earth.

In both cases, a hundred-fold decrease in intensity corresponds to an addition of 5 to M or m. Each time the magnitude increases by one, the brightness reduces by the same factor. While the original classification was discrete (stars could be first or second magnitude but not magnitude one-and-a-half), we now allow M and m to vary continuously.

(a) Show that a first magnitude ($m = 1.0$) star is $\sqrt[5]{100} = 2.512\ldots$ times brighter than a second magnitude ($m = 2.0$) star.

(b) Show that the brightness of a star (as seen from Earth) is proportional to $10^{-2m/5}$. Once you have showed this, it follows that the total visible power output of the star (its luminosity, in watts) must be proportional to $10^{-2M/5}$.

(c) Using the results of the last section, the intensity of a star when viewed from a distance D will be given by $I = P/4\pi D^2$, where P is the total visible power output (luminosity) of the star. Use your answers from (b) to show that this means that $10^{-2m/5} = B \times 10^{-2M/5}/D^2$ where B is a constant. Please note that we can measure D in any units we choose, as long as we adjust the constant B accordingly.

(d) The absolute magnitude M is defined to be the apparent magnitude the star would have if it were 10 parsecs (pc) away from Earth. Assuming that we measure our distances D in parsecs, show that the constant $B = 100$ pc².[†]

---

*That said, if you are working in 2-Ds, the energy does not spread out so rapidly. A ripple on a pond has circumference (hence ripple length) $2\pi r$, and therefore as the ripple gets bigger we expect power = intensity $\times 2\pi r$ to remain constant. Accordingly intensity $\propto 1/r$ and $E = E_0 \cos(\omega t - kr)/\sqrt{r}$.

[†]The parsec (pc) is about 3 light years, and is the distance from the Sun at which an observer would see the radius of the Earth's orbit subtending 1 second of arc (i.e. 1/3600°). Please see Q1 of Section 3.1.1.

(e) By taking logarithms to base ten of the result from (c), or otherwise, show that the distance D of a star from Earth is given, by $10^{(5+m-M)/5}$ pc. You may find it helpful to study workshop 7.2 if you are in need of revision on logarithms.

In practice, the distances of many stars in our own, or other, galaxies are calculated using this formula. The value of m is determined from Earth-bound observation, while M is inferred from other data known about the star – for example, its spectral class. Particularly accurate values of M can be obtained for stars called Cepheid variables, whose brightness varies periodically, and for a particular type of supernova.

4.12 Conclusion

We began the chapter by applying our knowledge of trigonometry to describing oscillatory motion. After revision of the ideas of complex numbers, we were able to describe oscillations in a new way, which was more convenient when it came to analysing the damped oscillator. Travelling waves were then described, and interference between two waves of different paths could be analysed.

The physics which gives rise to travelling waves was discussed using the transverse wave on a taut string as a specific case. This led us naturally on to impedance, and the efficiency of wave coupling at a junction. At the end of the chapter we noted how our equations changed if we were working in two or three dimensions.

5
Circuits

When asked at the dawn of the 20th century why Britain produced so many world leading physicists, the Nobel laureate and Cavendish Professor J.J. Thomson replied that it was because no science was taught in British schools.

Understanding an electric circuit is hardly as conceptually groundbreaking or fundamentally fascinating as more exotic topics such as quantum mechanics, particle physics or astrophysics. However, the humble electric circuit is important to the engineers or physicists for two main reasons. First of all, they spend much of their time using and designing electronic circuits to do their measuring and experimental control. Without measurement, there is no science; and in a modern paper mill, hospital, refinery, nuclear reactor, airliner or laboratory, most of the measurements are made electronically. Second, a good understanding of electric circuits is not easy to achieve, but serves as an excellent exercise for the mind – which is then much more adept at tackling other problems (just like Thomson's school studies in Latin and Greek). In fact when engineers are trying to understand a completely new system, they may often draw analogies between the parts of the system and the components of an electric circuit. Once they have solved the circuit using their electrical knowledge, they have solved the other system too. Electric circuits, with all their connections, serve as excellent analogies for a vast number of linked or connected systems.

5.1 Fundamentals

Electricity is a wonderful tool for transferring energy (and hence information) from one place to another. When a battery or generator is put into a circuit, it sets up electric fields. All of the charged particles involved experience forces as a result of this field. Electrons are repelled by the (−) terminal and attracted to the (+) terminal of the supply, the nuclei in the wires (and the insulators) experience forces in the opposite directions. Accordingly, when some of the electrons are free to move, as they typically are in a metal, they will accelerate. As they accelerate, they are involved more frequently in collisions which resist their motion, just as falling sky divers experience more resistance from the air as they fall faster. The sky divers eventually reach terminal velocity, and once this has been reached, their weight and the air resistance are equal and opposite, and they accelerate no more. Similarly, the electrons reach a certain top speed called their drift velocity with great rapidity, and can then be thought of as travelling round the circuit at a steady speed.

The analogy with the sky diver is far from perfect. All of the atoms in the sky diver, for example, fall with the same speed, since they are attached to one another. The electrons in the wire, on the other hand, have a great range of velocities up, down and across the wire (even when no battery is in the circuit) – with the fastest travelling at an appreciable fraction of the speed of light. However, when we consider the average (or mean) velocity of the electrons, the analogy is a fair one. Before the battery is put into the circuit, the mean velocity (or drift velocity) is zero. Once the battery is connected, each electron acquires a small additional component of velocity in the same direction, and the mean velocity increases by this same amount too.

5.1.1 Electric current

So, what exactly is carried round an electric circuit? The first answer is that *electrically charged particles* are pushed round in response to the fields. The electric fields in wires cause the free electrons to move, while the chemical reactions in cells, and the electromagnetic effects in generators have similar effects. Accordingly, we wish to measure this flow of charged particles. We call it the *current* and measure it with an ammeter. We define the current at a point in a circuit as the rate (in coulombs per second) and direction in which a stream of positively charged particles would have to flow to have the same electrical effect as the one actually observed.

In a metal wire, the charge carriers are electrons, and are negatively charged. If 10^{16} electrons pass a point in a wire each second (travelling from $(-)$ to $(+)$), then this is equivalent to 10^{16} oppositely charged particles travelling the other way each second. Since each of these 'oppositely charged particles' would carry a charge of $+1.6 \times 10^{-19}$ C, this means that we have a flow of $10^{16} \times 1.6 \times 10^{-19}$ C = 0.0016 C each second, so the current is 0.0016 A or 1.6 mA, and this current is said to flow from $(+)$ to $(-)$ in the wire.*

Q1 A metal wire of cross-sectional area A contains n free electrons per cubic metre. If all of these electrons move along the wire in the same direction at speed u, and each one has charge q, what is the current?†

Q2 Using your answer to Q1, work out the mean (or drift) velocity u for electrons carrying a steady 13 A current in a copper cable with a cross-sectional area of 1.5 mm². Assume that copper has 1.0×10^{29} free electrons per cubic metre. Does your answer surprise you?

Q3 Recalculate Q2 for a 0.1×0.2 mm cross section of silicon, with about 10^{20} free electrons per cubic metre, carrying a current of 10 mA. Does your answer surprise you?

---

*Note that this means that inside the battery (or generator), the current must be flowing the other way from $-$ to $+$. It is able to flow the 'wrong' way because it is being pulled by the chemical or electromagnetic forces present there, which are sufficiently great to overcome the electrical repulsion.

†*Hints*: (1) Work out the number of free electrons per metre of wire. If you look at a particular point on the wire, the electrons which are about to pass that point in the next second occupy a cylinder behind that point of length u. So, (2) work out the number of electrons which occupy a length u of wire, as this is the number of electrons which will pass a point in 1 s. Finally, (3) work out the total charge on these electrons. This gives the current.

In an ionic solution, copper II sulphate for example, there could be positive (Cu^{2+}) and negative (SO_4^{2-}) ions on the move. The copper ions would move one way, the sulphate ions the other. However, they would both contribute to carrying the electric current *the same way*.

Q4 A beaker of copper II sulphate is carrying a current of 30 mA between two electrodes. Assume that all of the current is carried by the copper ions. Calculate the number of copper ions leaving the anode ((+) electrode) each second. Also calculate how many free electrons leave the anode each second to flow round the rest of the circuit. How would the answers change if the same total current of 30 mA were carried equally by copper and sulphate ions? +

Notice that at the electrode, different numbers of charged particles can arrive as leave. At the cathode, copper ions arrive, and have their charge neutralized by the electrons arriving down the wires. No charged particles leave. However, the current arriving (on the positive ions) is equal to the current leaving (via the negative, arriving electrons).

5.1.2 Electric potential

So, what exactly is carried round an electric circuit? Our first answer was that charged particles (and hence electric charge) was carried round. However, the reason we spend so much time making electric circuits is that by means of that electric charge, energy (and hence information) can also be carried.

As mentioned in the introduction, the moving electrons gain energy from the electric field in the wires. The field is set up by, and gets its energy from, the processes in the cell or generator. However, the electrons reach a low 'terminal' velocity quickly, and so their kinetic energy never gets terribly high. The energy from the field, accordingly, is directly passed on to the other particles in the wire via collisions which the electrons make. This energy warms the wire and its surroundings. Other components, like light-emitting diodes (LEDs), motors or loudspeakers, are able to convert the energy in the electric field into light, motion or sound.

So while we cannot talk strictly of electrons 'carrying' electrical energy round the circuit,* it does make sense to say that between two points in the circuit, the charge carriers will take a particular amount of electrical energy from the field and convert it to other forms. This drop in electrical energy will of course be measured in joules, and when we work it out per unit of charge, it is called the potential (or voltage) difference between the two points. The point of higher potential is the one from which a positive charge would naturally be repelled towards the other; so the (+) terminal of a battery has a more positive potential than the (−) terminal.

Q5 Use your earlier answer to work out the kinetic energy of the electron in Q2. The potential difference of the mains supply is about 230 V. Calculate the potential

---

*The energy is in the field, not the electron (unless you count the negligible amount of kinetic energy the electron has, which is constant all the way round the circuit).

energy lost by the electron as it goes through a 230 V light bulb. What is the ratio of the potential to kinetic energy?

While you can measure the potential difference across any component, or between any two points in the circuit, it is frequently more helpful to measure the potentials of all the points in the circuit with reference to a single point. Usually the (−) terminal of the battery is chosen as the reference point. That way, the potentials of all the other points are higher and therefore regarded as positive. The potential of the reference point (measured with respect to itself) is of course zero.

The labelling of points with voltages can be a helpful tool in analysing circuits. We shall use it here to solve a simple problem – working out which bulbs in a circuit are going to be bright and which dim.

5.1.3 Workshop: Using voltage to solve simple circuit problems

For each circuit in Figure 5.1:

- Label the negative terminal of the battery (top left point) '0 V'.
- Label the other points on the upper side of the circuit with their correct voltages. For each cell, the (+) terminal's voltage will be 1.5 V higher than its (−) terminal.
- Now label the other points on the circuit with their voltages. Assume that the voltage of the two ends of the same wire will be the same, as very little energy is converted to heat by a good conductor unless the currents are really high.
- Any bulb with exactly 1.5 V difference across it will be normal brightness (N). If the potential difference is greater, the bulb will be brighter (B); if it is less, the bulb will be dim (D) or off (O) if the potential difference is zero. Label each bulb with a B, N, D or O according to its brightness.

5.1.4 Ohm's law and resistance

When a component like a bulb or heater is supplied with a larger voltage, a stronger electric field is set up. Notwithstanding the collisions, you would expect the stronger electric field to cause a higher drift velocity for the electrons just as you would expect terminal velocities of falling objects to be higher if the gravitational field strength got larger.* Accordingly, when the voltage goes up, the electrons move faster, and a larger current is carried. For many materials, the voltage and current are proportional, and we say that these materials obey Ohm's law and so we find that the ratio

$$R = \frac{V}{I} \tag{5.1}$$

is constant. This ratio is called the resistance, and it is measured in ohms (Ω). Other materials do not obey Ohm's law, but we still define their resistance using equation (5.1) – but the resistance will be a function of the current through the component.

Notice that because the voltage V measures (in some way) the strength of the electric field (and hence the force on a charge carrier), and I measures the current

---

*That is, assuming that the atmospheric density remained the same (which it would not).

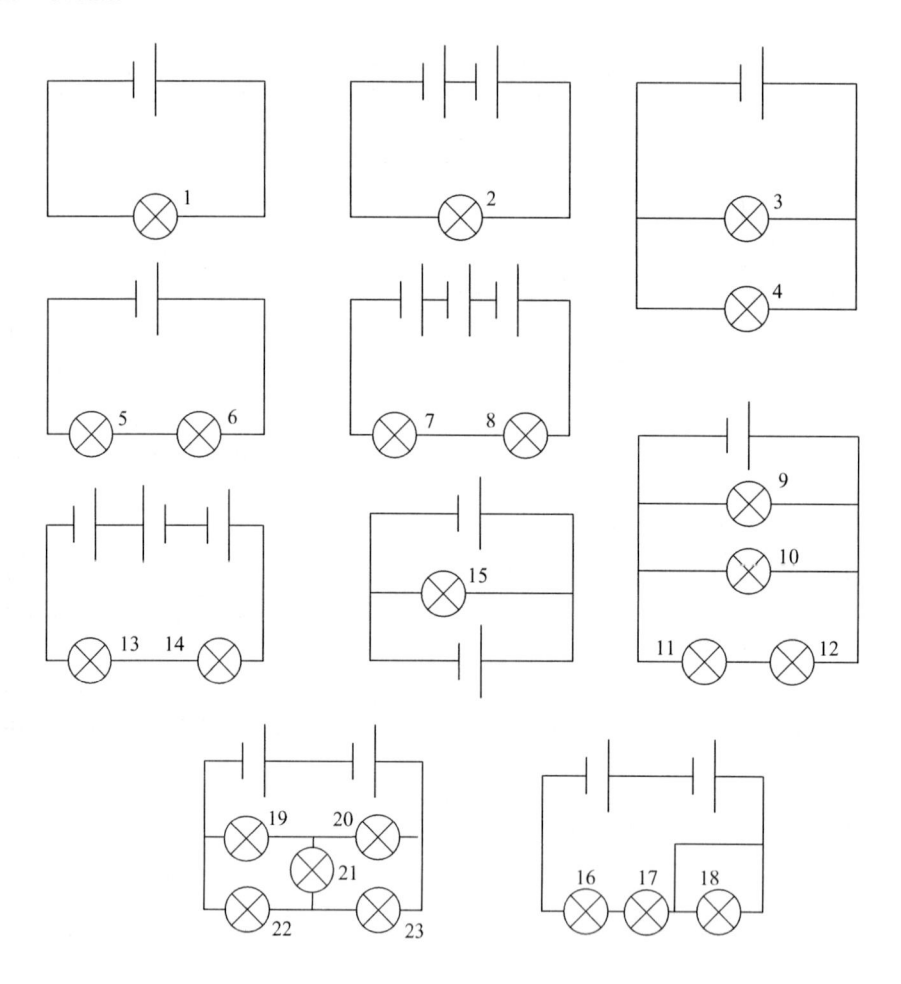

Fig. 5.1

and hence the speed of the carriers, it follows that $R = V/I$ measures a kind of 'force per unit speed'. Given the discussion at the end of Section 4.9, where the impedance of a medium to the motion of a wave was equal to the ratio of the force to the speed, it should come as little surprise that R behaves in many ways like an impedance. We shall be exploring this link further later in this chapter.

5.2 Direct current circuit analysis

While it is helpful to review the processes and terminology at the heart of an electric circuit, part of our motivation for studying circuits was to gain a way of understanding connected systems. Therefore we turn from electrons and potential energy to whole circuits. We start with the simplest kind of circuits – ones with perfect batteries, perfect resistors and perfect wires to connect them up.

Direct current circuit analysis 5.2 **103**

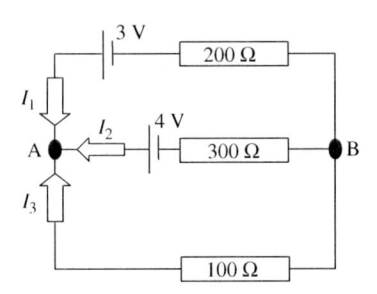

Fig. 5.2

- A perfect wire has no resistance, so all of its points will be at the same voltage (potential).
- A perfect resistor has a constant value of resistance R, such that whatever the voltage V put across it, the current flowing through it will be given by $I = V/R$.
- A perfect 3 V battery always has a 3 V difference between its $(+)$ and $(-)$ connections, irrespective of how much current is flowing through it – in other words, the perfect battery has zero resistance.

That said, the circuits can be quite complicated, because they might be connected in very complicated ways. Before we deal with a really complicated circuit, let us study Figure 5.2.

5.2.1 Analysis using fundamental principles

We want to find out the currents flowing through each resistor. To do this, we make use of two laws – Kirchhoff's laws.

The first law says that the total current leaving a junction must equal the total current entering it. To give an example, if an 80 A current flows into a domestic fusebox, which is connected to an oven drawing 38 A and some sockets, then the sockets must be drawing 80 A $-$ 38 A $=$ 42 A.

In our example, that means that $I_1 + I_2 + I_3 = 0$.

Note that it does not matter which way the arrows are drawn on the diagram to start with. If one of them is drawn the 'wrong' way, we shall simply get a negative answer for that current, telling us that it is flowing in the opposite direction to the arrow.

The second law says that you must get the same answer for the voltage difference between two points whichever route you calculate it from.

In our example, let us use V to mean the voltage of point A relative to B. It follows that

$$V = 3\text{ V} - 200\,I_1 \quad \text{using the top branch of the circuit}$$
$$V = 4\text{ V} - 300\,I_2 \quad \text{using the middle branch}$$
$$V = -100\,I_3 \quad \text{using the lowest branch.}$$

Notice that we add the voltage of a cell when going through it from (−) to (+). This is, after all, the direction the current goes through the cell, and the charge gains energy as it does so – so the voltage must increase*.

Notice that we *subtract* the voltage across a resistor when going through it in the direction of the current. This is because as the charge passes through the resistor in this direction it is losing electrical energy, so its voltage must be decreasing†.

We now have four equations in our four unknown quantities V, I_1, I_2, and I_3, and these can be solved to tell us that the voltage of point A relative to B is about 1.55 V.

Q6 Solve the circuit, and check that $V = 1.55$ V. Also find the currents through the three resistors. Which current or currents are going in opposition to the arrows on the diagram?

5.2.2 Method of loop currents

Complex circuits are better solved more systematically. One approach is to break them down into their smallest loops, and assign a current to each loop, as in Figure 5.3. Current I_3 is in the second loop, so $I_3 = I_{L2}$. Current I_1 is in the first loop, but in the opposite direction to the loop, so $I_1 = -I_{L1}$. Other components (such as the 300 Ω resistor in our example) are in both loops, and Kirchhoff's first law can be used to say that the current in this resistor is $I_2 = I_{L1} - I_{L2}$.

We then form one statement of Kirchhoff's second law for each loop. If you go all the way around a loop, the total potential difference must be zero (so that you come back to the same voltage you started with). In our circuit, it means that if you start at point A and go round the loops clockwise, our two voltage loop equations are
First loop:
$$-3 - 200\, I_{L1} - 300(I_{L1} - I_{L2}) + 4 = 0$$
so
$$1 = 500\, I_{L1} - 300\, I_{L2}$$

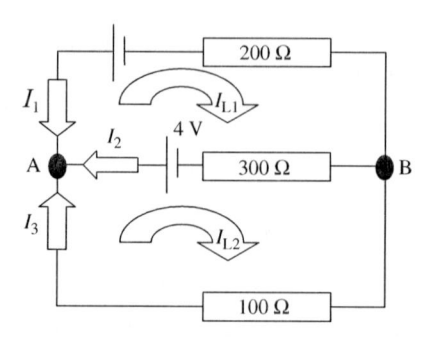

Fig. 5.3

---

*If going through a cell in the opposite direction to the current, we subtract the voltage.
†If going through a resistor in the opposite direction to the current, we add the voltage.

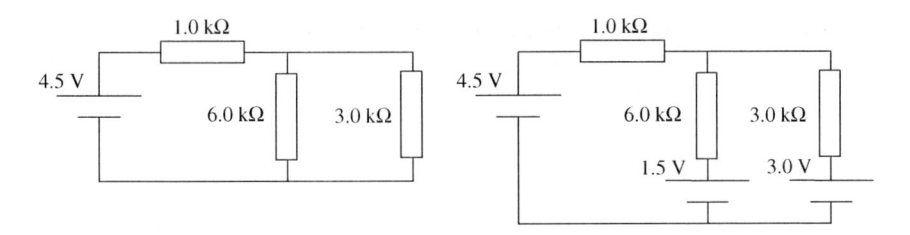

Fig. 5.4

Second loop:
$$-4 - 300(I_{L2} - I_{L1}) - 100\,I_{L2} = 0$$

so
$$4 = 300\,I_{L1} - 400\,I_{L2},$$

and so we have two equations in two unknowns I_{L1} and I_{L2}, which can be solved to give $I_{L1} = -7.27$ mA while $I_{L2} = -15.45$ mA (to two decimal places).

Notice that the loop method is simpler in that there were two simultaneous equations rather than four, and it is more easy to generalize it to more complex networks.

Q7 Calculate the current in each resistor in the networks in Figure 5.4.

5.3 Introducing alternating current

In alternating current (a.c.) analysis, instead of working with perfect batteries, we work with perfect alternators – that is, sources of perfect sine wave signals. This analysis is useful because it enables us to study our mains electricity supply and also enables us to study any oscillating electrical signal. These signals could be anything from the output of a microphone which is amplified in a mixer to the oscillating microwaves used to convey information to a communications satellite. As such, a.c. analysis is valuable in helping us understand all kinds of communication or signalling.

If the voltage across a component is given by $V = V_0 \cos \omega t$, then after any short-term transients have died away, we find that the current is also given by a cosine wave: $I = I_0 \cos(\omega t + \phi)$. As we can see from the presence of ϕ this wave might not be in step with the voltage. When solving the circuit, we aim to find out two things: how large I_0 is in relation to V_0 and the size of the phase factor ϕ.

It turns out to be more simple mathematically if we use the complex exponential notation for these oscillations.* We may then write

$$V = \mathrm{Re}(V_0 e^{i\omega t}), \tag{5.2}$$

$$I = \mathrm{Re}(I_0 e^{i(\omega t + \phi)}) = \mathrm{Re}(I_0 e^{i\phi} e^{i\omega t}). \tag{5.3}$$

---

*If you are rusty on complex numbers, please work through Section 4.2 before proceeding. If you are used to complex numbers, but have not seen them used to represent oscillations before, please read Section 4.3.

If we write the *complex amplitude* of the current as $J_0 = I_0 e^{i\phi}$, then things are simpler still, since the one complex number J_0 contains both the amplitude and phase information.

The impedance of the component is defined as the complex number $Z = V_0/J_0$. The modulus of Z gives the ratio of the amplitudes of the voltage and current waves, while $\text{Arg}(Z)$ gives the phase difference between the two waves.

You will frequently see the impedance written as the sum of a real part (called the resistance R) and an imaginary part (called the reactance X). Accordingly, $Z = R + iX$, where R and X are real numbers.

5.3.1 Resistors

The resistor is the simplest kind of component. For the resistor, the voltage V and current I are proportional: $V = IR$. Accordingly, if $I = I_0 \cos \omega t$, then $V = I_0 R \cos \omega t$. The phase factor between voltage and current waves is accordingly zero, and $Z = V/I = R$ is a real number. Now you can see that the real part of impedance is indeed resistance.

Q8 A component has an impedance of 18 Ω. It is connected to an a.c. electrical supply with frequency 50 Hz and amplitude 325 V.
  (a) Calculate the amplitude of the current.
  (b) Calculate the amplitude of the electron speed in the connecting wire. Assume that the wire is made of copper (with 10^{29} electrons/m^3) and has cross-sectional area 1.5 mm^2. (*Hint*: Use your answers to Q1 or Q2.)
  (c) Calculate the amplitude of the electron's motion in the wire as it oscillates back and forth. You may find equation (4.9) helpful.

5.3.2 Power in a.c. circuits and rms values

If the voltage across a component gives the potential energy difference per unit charge between the charges going in and those coming out, and the current gives the charge passing through the component every second, then it makes sense that current multiplied by voltage gives the electrical potential energy lost each second – that is, the power transformed from electrical energy to other forms by that component.

For a direct current (d.c.) circuit, we sum this up as $P = IV$. Here P is regarded as positive when the component is using electrical energy, and V is regarded as positive when the potential goes down as the charges pass through the component in the 'positive I' direction. If the potential rises as the charge goes through, then either V or I is negative, and so P is negative too. This implies that the component is *supplying* electrical energy to the circuit.

When working with a.c., there is a refinement we need to make. Since V and I are not steady, P will not be steady either. The power supplied to a resistor is given by

$$\begin{aligned} P &= VI \\ &= V_0 \cos \omega t \, I_0 \cos \omega t \\ &= V_0 I_0 \cos^2 \omega t. \end{aligned} \quad (5.4)$$

Accordingly, the average value of the power is $\overline{P} = \frac{1}{2}V_0 I_0$.* As the component is a resistor then $V_0 = RI_0$, and we can also write the average power as

$$\overline{P} = \frac{1}{2}I_0^2 R = \frac{\frac{1}{2}V_0^2}{R}.$$

These factors of one-half would make life confusing. They would mean that while $P = IV$ for d.c. circuits, $P = \frac{1}{2}IV$ for a.c. However, if we describe the alternating current and voltage not by their amplitudes but by $V_{\text{rms}} = \frac{1}{\sqrt{2}}V_0$ and $I_{\text{rms}} = \frac{1}{\sqrt{2}}I_0$, then we notice that $\overline{P} = I_{\text{rms}} V_{\text{rms}} = I_{\text{rms}}^2 R = V_{\text{rms}}^2/R$, and the formulae are identical to those for d.c. You could say that V_{rms} (or I_{rms}) is the steady voltage (or current) which would heat a resistor identically to the a.c. actually present. In fact these values are so useful that whenever an a.c. measurement is taken, you can assume that it is the rms value that is meant unless told explicitly otherwise.

Q9   For the British mains supply V_{rms} is supposed to be close to 230 V. Calculate the amplitude of the voltage.

Q10 Calculate the maximum current going through a heater if $I_{\text{rms}} = 13$ A.

Q11 What is the maximum power consumption of a '60 W' light bulb?

Q12 The letters 'rms' stand for 'root mean square'. You can understand why this description is used as follows:
(a) Suppose $V = V_0 \cos \omega t$. Calculate V^2.
(b) Calculate the mean value of V^2.
(c) Now take the square root. You should get $\frac{1}{\sqrt{2}}V_0$.

Notice that you have worked out $\sqrt{\overline{V^2}}$, that is the root of the mean of the square, or rms for short.

Q13 You may have noticed that in this question we refused to use complex exponential notation. The reason is that when you multiply a complex current by a complex voltage you must explicitly remember that the voltages and currents are only the real parts of $V_0 e^{i\omega t}$ or $I_0 e^{i\omega t}$, not the whole complex numbers. +
(a) Multiply $V_0 e^{i\omega t}$ and $I_0 e^{i\omega t}$ together. Notice that the result's real part is equally often positive and negative with a mean of zero. The true expression for power (5.4) on the other hand is never negative and has a mean of $\frac{1}{2}V_0 I_0$.
(b) To do the calculation properly, we use the result from Section 4.2(i) that $\text{Re}(z) = \frac{1}{2}(z + z^*)$. We shall also take the opportunity of avoiding the assumption that V and I are in phase. Using this formula, show that

$$VI = \text{Re}\left(V_0 e^{i\omega t}\right) \times \text{Re}\left(I_0 e^{i(\omega t + \phi)}\right) = \frac{1}{4}V_0 I_0 \left(e^{2i\omega t + i\phi} + e^{-2i\omega t - i\phi} + e^{i\phi} + e^{-i\phi}\right)$$

if V_0 and I_0 are real.

---

*We have assumed that the voltage and current are in phase, as this is a resistor. The situation is more complex if the two wave forms are not in phase, and the situation is tackled in Q13.

108 *Circuits*

(c) Using our result from Section 4.2(p) that $\cos\theta = \frac{1}{2}\left(e^{i\theta} + e^{-i\theta}\right)$, show that $VI = \frac{1}{2}V_0 I_0 \left(\cos(2\omega t + \phi) + \cos\phi\right)$, and hence that the average power is equal to $\frac{1}{2}V_0 I_0 \cos\phi = V_{\text{rms}} I_{\text{rms}} \cos\phi$.

(d) To reconvince you that complex numbers are better than sines and cosines, derive this result using trigonometry, where unlike in (5.4) above, you include the phase factor and assume that $I = I_0 \cos(\omega t + \phi)$.

The $\cos\phi$ term which has cropped up in our equations is called the power factor. You can see that if the voltage and current get out of phase, less power can be transmitted for the same maximum voltage and current. For this reason, electricity supply companies charge extra to commercial customers whose equipment causes the two waves to get out of phase. Domestic electricity meters are not designed to pick this up, which these days saves many customers' money at the generator's expense – modern 'energy efficient' devices such as compact fluorescent light bulbs and solid state power supplies (as used by laptops) frequently have power factors as low as 0.5.

5.3.3 Capacitors

The voltage across a capacitor is given by Q/C where Q is the charge stored on it,[*] and C is its capacitance. If a charging current I flows through the capacitor, then $I = dQ/dt$. It follows that the fundamental equation for a capacitor is

$$I = \frac{dQ}{dt} = \frac{d}{dt}CV = C\frac{dV}{dt}. \tag{5.5}$$

Accordingly, if $V = \text{Re}(V_0 e^{i\omega t})$, then

$$I = C\frac{dV}{dt} = C\frac{d}{dt}\text{Re}\left(V_0 e^{i\omega t}\right)$$
$$= \text{Re}\left(CV_0 \frac{de^{i\omega t}}{dt}\right) = \text{Re}\left(i\omega C V_0 e^{i\omega t}\right).$$

The complex amplitude of the current $J_0 = i\omega C V_0$, and so the impedance, which is defined as $Z = V_0/J_0 = 1/i\omega C = -i/\omega C$.

The phase difference ϕ is calculated from $\text{Arg}(Z) = -\pi/2$.[†] Accordingly, the current peaks one-fourth of a cycle before the voltage. Engineers talk of the 'current leading the voltage' in a capacitor.

Using our answer to Q13, we can see that with this value of ϕ, the power factor $\cos\phi = 0$, and accordingly no power is dissipated in the capacitor on average. At some times during the cycle, the instantaneous power $VI < 0$. At these times the capacitor is charging and the energy stored in its electric field is increasing. At other times

---

[*] As is conventional, when we say charge Q is stored on the capacitor, we mean that $+Q$ is stored on the positive plate, and $-Q$ is stored on the negative plate. The total on the two plates is always equal to zero.

[†] See part (h) of workshop 4.2 for an explanation of what we mean by 'argument'.

$VI > 0$. At these times the capacitor is discharging, and energy is being transferred from the capacitor's electric field to the rest of the circuit.

5.3.4 Inductors

Inductors are usually coils of wire which form a magnetic field when a current passes. The total number of magnetic field lines Φ will be proportional to the current, so $\Phi = LI$, and we call the constant of proportionality the inductance L. When the current changes, the magnetic field changes and this induces a voltage in the coil. This voltage can be calculated from Faraday's law as the rate of change of magnetic field $V = d\Phi/dt$, so $V = L\, dI/dt$.

Accordingly, if $I = \mathrm{Re}(I_0 e^{i\omega t})$, then $V = \mathrm{Re}(i\omega L I_0 e^{i\omega t})$. This means that $Z = i\omega L$.

The phase difference ϕ is now $+\pi/2$, so the current peaks one-fourth of a cycle after the voltage. Engineers talk of the current in an inductor 'lagging' the voltage.

As with the capacitor, a perfect inductor has an average power consumption of zero because the voltage and current are $\pi/2$ radians out of phase.

5.3.5 Sign conventions

When we write $V = IR$ for a resistor, the V stands for the size of potential difference (or voltage drop) across it, while I stands for the size of current through it. As such, no sign is needed. However, when we analyse a circuit, we need to give the directionality some thought. As mentioned earlier, the voltage drops (as electrical energy is lost) in the direction the current flows.

This same convention works for the capacitor. If I is positive and charging the capacitor, the current is flowing onto the $(+)$ plate, and so similarly the voltage drops in the direction of positive current flow.

For the inductor if the current is increasing, the voltage will be induced so as to oppose this change. So the voltage drop is in the direction of the current if $dI/dt > 0$, and vice versa. Accordingly we may write $V = L\, dI/dt$ where V is regarded as positive if it goes down in the direction of positive I.*

To sum up: all of our equations relating the voltage difference across a component to the current through it follow the same sign convention – a positive value of V means that the potential is higher at the end where a positive current goes in and lower at the end where a positive current comes out as summarized in Figure 5.5.

5.3.6 Phasor methods in a.c. analysis

While complex analysis provides the quickest and cleanest way of solving problems in a.c. analysis, the more visually minded among us have always used 'phasor' diagrams to aid understanding of this topic. The diagram is basically a drawing of the complex

---

*Many books take a different view, and insist on labelling the voltage across an inductor as $-L\, dI/dt$, to make it clear that the voltage drops as you go across the component when a current is increasing. However, if they were being consistent, they would also use $V = -IR$ for the humble resistor since the voltage also drops as you go through a resistor in the direction of the current. You are welcome to put in the minus signs if you wish, however, you are going to be very confused if you use a minus sign for the inductor and not for the resistor.

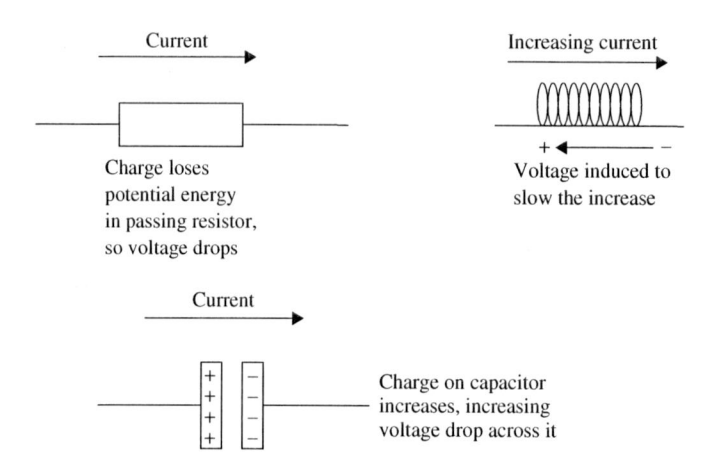

Fig. 5.5

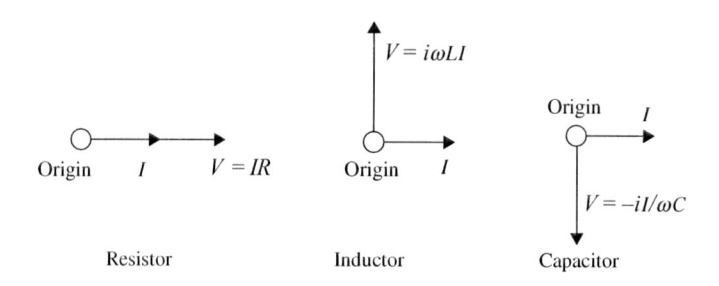

Fig. 5.6

amplitudes of V and I in the complex plane or Argand diagram of Section 4.2. The voltages and currents are represented as 'vectors' in the complex plane with lengths given by their amplitudes and directions representing their phase. As the current in the circuit goes back and forth, the phasor 'vectors' rotate in a manner analogous to the vectors used for describing rotation in Section 3.1.

The diagrams showing the relationship between current and voltage for our three main components are given in Figure 5.6.

If you imagine all of the arrows rotating anticlockwise at the frequency of the a.c., you can see the meaning of the current 'lagging' the voltage for the inductor or 'leading' it in the case of the capacitor.

Many electrical engineers or electricians will use the arrows on these diagrams like vectors when they analyse circuits. In a series circuit, the current through the components will all be the same, and so one I arrow is drawn pointing to the right. The 'voltage' arrows are drawn for each component in turn and then added vectorially to give the voltage across the combination. The phase of the resultant voltage with respect to the current can then be calculated from its direction, and its amplitude is given by its length.

When working with parallel circuits, the electrician rotates each of the figures above until the voltage arrow points to the right, since the voltage will be the same for each element in the circuit. The currents through each component are then drawn in the correct directions* and added vectorially to give the total current through the combination.

5.4 Alternating current circuit analysis

We now extend the methods used for d.c. circuit analysis in Section 5.2 to a.c.. Our first circuit is shown in Figure 5.7 – an inductance L wired in series with a resistance R. This is a useful circuit, since an electric motor can be modelled well as a resistor and inductance in series.

This circuit has only one loop, and we denote the current in this loop as I. Kirchhoff's second law then gives (starting at the black dot and working clockwise):

$$V_0 \cos \omega t - IR - L\, dI/dt = 0. \tag{5.6}$$

The general solution to this equation contains two parts – one which dies away with time and one (of frequency ω) which remains. This remaining part is called the 'steady state' current (even though it is oscillatory). If the circuit is in use for some time, we can neglect the part which dies away, and we can use our understanding of impedance to solve the problem.

$$\mathrm{Re}(V_0 e^{i\omega t}) = \mathrm{Re}(RJ_0 e^{i\omega t} + i\omega L J_0 e^{i\omega t}). \tag{5.7}$$

If we write down the more restrictive equation:

$$V_0 e^{i\omega t} = RJ_0 e^{i\omega t} + i\omega L J_0 e^{i\omega t}, \tag{5.8}$$

then any value of J_0 which satisfies this will also automatically satisfy our equation (5.7). The value is found as

$$J_0 = \frac{V_0}{R + i\omega L}. \tag{5.9}$$

Notice that the impedance of this resistor–inductor circuit ($Z = V_0/J_0$) is equal to $R + i\omega L$. In other words, the impedance of a series circuit is equal to the sum of the impedances of its parts.

The current therefore has an amplitude given by

$$I_0 = |J_0| = V_0/\sqrt{R^2 + \omega^2 L^2}, \tag{5.10}$$

and the phase of the voltage with respect to the current is given by

$$\mathrm{Arg}(Z) = \mathrm{Arg}(R + i\omega L) = \tan^{-1}(\omega L/R). \tag{5.11}$$

---

*That is, right for a resistor, down for an inductor and up for a capacitor.

112 Circuits

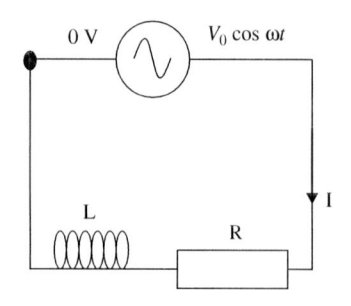

Fig. 5.7

You can also find the phase of the current relative to the voltage if we write the impedance of the circuit as $|Z|e^{i\phi}$ where $\phi = \text{Arg}(Z)$.

$$J_0 = \frac{V_0}{|Z|e^{i\phi}} = \frac{V_0\, e^{-i\phi}}{|Z|}, \tag{5.12}$$

and so the phase of current with respect to voltage is $-\phi = -\text{Arg}(Z)$.

In one sense this is obvious – if the phase of voltage with respect to current is $+\phi = \text{Arg}(V_0/J_0)$, then the phase of current with respect to voltage must be $-\phi = \text{Arg}(J_0/V_0)$. However, notice that how well complex numbers are suited to this kind of algebra – since with complex numbers $1/Z$ always has the opposite argument to Z.

5.4.1 Analysis using impedances

We have deliberately 'laboured' this example to show the use of the complex numbers and the application of Kirchhoff's second law. However, when there is only one power supply, it is easier to analyse the circuit purely in terms of the impedances.

Suppose that we have two components which have impedances of Z_1 and Z_2. If we connect them to a voltage $Ve^{i\omega t}$ (we omit the 'Re' for brevity), then we can work out the current flowing.

If the two are connected in series, then they will have the same current $Je^{i\omega t}$, and the supply voltage must be given by the sum of the voltages across the two components. Therefore,

$$\begin{aligned}Ve^{i\omega t} &= V_1 + V_2 \\ &= Z_1 Je^{i\omega t} + Z_2 Je^{i\omega t} \\ &= (Z_1 + Z_2)\, Je^{i\omega t}, \end{aligned} \tag{5.13}$$

and the overall impedance of the circuit is

$$\text{Series:} \quad Z = \frac{V}{J} = Z_1 + Z_2. \tag{5.14}$$

Alternatively, if the two are connected in parallel, they must share the same voltage, and the supply current $Je^{i\omega t}$ must be the sum of the currents through the

Alternating current circuit analysis 5.4

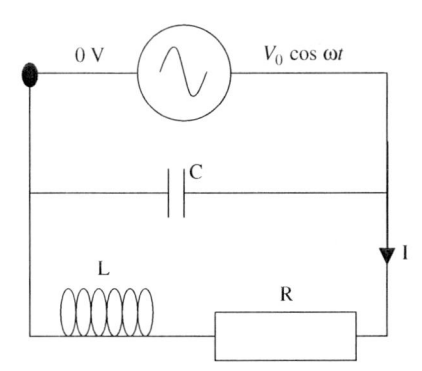

Fig. 5.8

two components. Therefore,

$$Je^{i\omega t} = J_1 + J_2$$
$$= \frac{Ve^{i\omega t}}{Z_1} + \frac{Ve^{i\omega t}}{Z_2}$$
$$= \left(\frac{1}{Z_1} + \frac{1}{Z_2}\right)Ve^{i\omega t}, \quad (5.15)$$

and the overall impedance of the circuit is given by

Parallel: $$Z = \frac{V}{J} = \left(\frac{1}{Z_1} + \frac{1}{Z_2}\right)^{-1}. \quad (5.16)$$

We now use these facts to work out the impedance of the circuit in Figure 5.8 – our old resistor–inductor circuit with a capacitor in parallel.

The impedance of the resistor–inductor branch is $R + i\omega L$, as we showed before. The impedance of the capacitor is $-i/\omega C$. Accordingly, the impedance of the combination is given by

$$Z = \left(\frac{1}{R + i\omega L} + \frac{1}{-i/\omega C}\right)^{-1}$$
$$= \left(\frac{1}{R + i\omega L} + i\omega C\right)^{-1}$$
$$= \frac{R + i\omega L}{1 + i\omega C(R + i\omega L)}$$
$$= \frac{R + i\omega L}{1 + i\omega CR - \omega^2 CL}. \quad (5.17)$$

Q14 Use equation (5.17) to derive the capacitance needed if the supply current is to be in phase with the supply voltage. Give your answer in terms of R, L, and ω.

114   *Circuits*

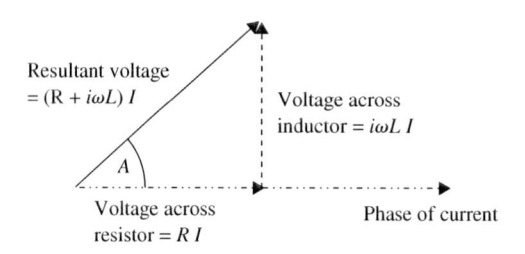

Fig. 5.9

Hint: To make the voltage and current to be in phase, the impedance must be entirely real. This means that Z^{-1} must also be real. In practice this is an important calculation. Many industrial plants contain motors, which have considerable inductance. This causes their currents to lag significantly behind the mains voltage, and this gives rise to financial penalties and wasted resources in the supply network as indicated in Q13. Installing capacitors in parallel with the motors corrects the problem and saves the companies money. +

Q15 The tuner in a radio uses a circuit very similar to the one in this section, with a very low resistance R. Show that there is one frequency at which the impedance is extremely high, and express this frequency in terms of L and C. In a radio, the signal from the aerial is passed through an amplifier and then on to our LCR circuit wired in parallel with the rectifier which feeds the audio amplifiers and thus the speakers. Assuming that the signal from the aerial contains radio waves of many different frequencies, explain why only the chosen one with its special frequency will drive measurable currents through the rectifier and thus be 'heard'. On old or cheap radios, the capacitance can be changed using the tuning knob to 'tune' the radio to the frequency of the radio station desired. Modern radios are digital and thus far more complicated. +

5.4.2 Analysis using a phasor

We start with the inductor and resistor in series, each carrying current I. The voltages are given by $i\omega L I$ and RI, respectively, and if we add these vectorially, we get the situation shown in Figure 5.9.

From Figure 5.9 we can see that the magnitude of the voltage will be $\sqrt{R^2+\omega^2 L^2}\times I$ and the voltage will 'lead' the current by phase angle, $A = \tan^{-1}\omega L/R$. This agrees with our analysis in equations (5.10) and (5.11).

If we now put our capacitor C in parallel with this combination, the voltage across C must be the same as that across our LR combination, but the current through it will be different. The two currents can be added to give the total current as shown in Figure 5.10.

Denoting the amplitude of the current through the resistor and inductor as $I_{\text{LR}} = V/\sqrt{R^2 + \omega^2 L^2}$, we can see that the resultant current will be in phase with the voltage

Phase of voltage

Resultant current *I*

Current through C
= *iω*C *V*

Current through L and R

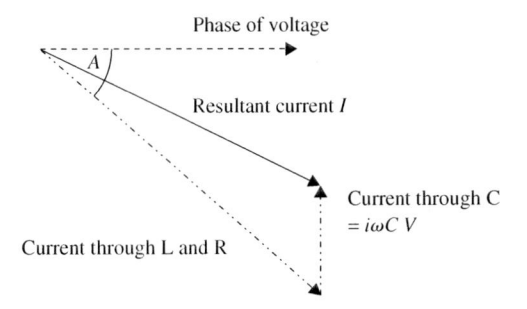

Fig. 5.10

if $I_{\text{LR}} \sin A = \omega \text{CV}$, and that the value of the capacitance needed to do this can be calculated from our calculations of A and I_{LR}.

Q16 Verify that the value of the capacitance needed to bring the current back in phase with the voltage agrees with your answer to Q14.

5.5 Conclusion

We began our chapter by presenting an overview of what is happening in terms of charge and energy in an electric circuit. We then used these principles to analyse d.c. circuit networks. Next we were able to use our knowledge of fields and waves (using complex numbers and phasors) to develop methods for analysing a.c. circuits with a single frequency.

6
Thermal physics

This chapter gives an introduction to the areas of physics known as thermodynamics and statistical mechanics. These deal with the questions, 'What happens when things heat up or cool down?' and 'Why?', respectively.

We start with a statement that will be very familiar – but then find that it leads us into new territory when explored further.

6.1 The conservation of energy: The first law

You will be used to the idea that energy can neither be used up nor created – only transferred from one object to another, perhaps in different forms.

For our purposes, this is stated mathematically as

$$\Delta Q + \Delta W = \Delta U, \tag{6.1}$$

where 'ΔX' refers to 'a change in X'. Put into words, this states: 'Heat entering object + work done on the object = the change in its internal energy'. Internal energy means any form of stored energy in the object. Usually this will mean the heat it has, which by the presence of matter, can be measured by temperature. However, if magnetic or electric fields are involved, U can also refer to electrical or magnetic potential energy.

Given that the conservation of energy must be the starting point for a study of heat, it is called the first law of thermodynamics.

Equation (6.1) can be applied to any object or substance. The most straightforward material to think about is a perfect or ideal gas, and so we shall start there. It is possible to generalize our observations to other materials and systems afterwards.

Imagine some ideal gas in a cylinder with a piston of cross-sectional area A. The gas will have a volume V, a pressure p, and a temperature T. From more elementary work (we also derive this result in Section 6.5.2) we remind ourselves that the internal energy of a monatomic ideal gas is related to its temperature by

$$U = \sum_i \frac{1}{2} m v_i^2 = \frac{3}{2} NkT, \tag{6.2}$$

where N is the number of molecules in the gas and k is Boltzmann's constant, which is about 1.38×10^{-23} J/K. We deal in more detail with perfect gases in Section 6.6, here we are only using the concept to help us visualize things. Such a gas is imagined to be made up of N identical tiny elastic spheres (so all the m are the same) in random motion (hence the different v_i), which do not interact with each other or the container, except through collisions with each other and the piston and cylinder. Therefore, the only energy they can be said to possess is kinetic and this total kinetic energy makes up the internal energy.

There are a number of terms that now need to be introduced to facilitate our discussions.

Adiabatic process: This is a process performed on our system that does not involve any heat energy being transferred in or out of our system. An easy way to represent this would be to say that $\Delta Q = 0$ in (6.1) for an adiabatic process.

Isothermal process: This is a process that occurs whilst maintaining a constant temperature in our system. Our system is an ideal gas so an isothermal process would be represented as $\Delta U = 0$ in (6.1).

Isochoric process: This is a process that involves no change in volume. For our system this would be like locking the piston position and could be represented by $\Delta W = 0$ in (6.1), since the only way to do mechanical work on the gas is to move the piston.

Let us now do some work on the gas by pushing the piston in by a small distance* Δx. The force F required to push the piston is $F = pA$, where A is the cross-sectional area of the piston. The work done on the gas is

$$\Delta W = F \Delta x = pA \Delta x. \tag{6.3}$$

Notice that $A\Delta x$ is also the amount by which the volume of the gas has been *decreased*. If ΔV represents the change in volume then $\Delta V = A \Delta x$. Therefore,

$$\Delta W = -p \, \Delta V.$$

For a perfect gas in a cylinder (or in fact in any other situation), the first law can be written a bit differently as

$$\Delta U = \Delta Q - p \, \Delta V. \tag{6.4}$$

6.2 The second law

While the first law is useful, there are certain things it can never tell us. For example, think about an ice cube sitting on a dish in an oven. We know what happens next – the ice cube melts as heat flows from oven to ice, warming it up until it reaches melting point. However, the first law does not tell us that. As far as it is concerned it is just as possible for heat to flow from the ice to the oven, cooling the ice and heating the oven.

We have stumbled across our next law – called the second law of thermodynamics. This can be stated in several ways, but we shall start with this: *Heat will never flow from a cold object to a hotter object by itself.*

---

*The work done is of course a scalar product of the force and displacement vectors. However, the force and displacement are clearly parallel for the piston so we have dropped the vector notation here and refer only to the scalar magnitudes.

118 Thermal physics

This helps us with the ice in the oven, but you may be wondering what the significance of the 'by itself' is. Actually heat *can* be transferred from a cold object to a hotter one – that is what fridges and air conditioning units do. However, they can only do it because they are plugged into the electricity supply. If you are prepared to do some work – then you can get heat out of a cold object and into a hotter one, but as soon as you turn the power off and leave it to its own devices, the heat will start flowing the other way again.

6.3 Carnot's theorem

6.3.1 Heat engines and fridges

The fridge is shown in Figure 6.1. It is a device which uses work ΔW (usually provided by an electric compressor) to extract heat ΔQ_2 from the icebox, and pump it out into the surroundings. We usually employ the word 'reservoirs' to describe the icebox and surroundings here, because we mean that they are large enough so that taking heat out of them or putting heat into them will not change their temperatures appreciably. The temperature T_1 of the surroundings is of course greater than the temperature of the icebox T_2. However, by the conservation of energy, the amount of energy pumped out ΔQ_1 is bigger than the amount of energy removed from the icebox. By convention $\Delta Q_2 > 0$ and $\Delta Q_1 < 0$, since heat flowing into this engine is regarded as positive. The first law therefore states, for this engine, that $\Delta Q_1 + \Delta Q_2 + \Delta W = 0$.

The fridge is a device that uses work to move heat from cold objects to hot. The opposite of a fridge is a heat engine (Figure 6.2). This allows heat to flow in its preferred way – namely from hot to cold – but arranges it to do some work on the way. Petrol engines, steam engines, turbogenerators, and jet engines are all examples of heat engines.

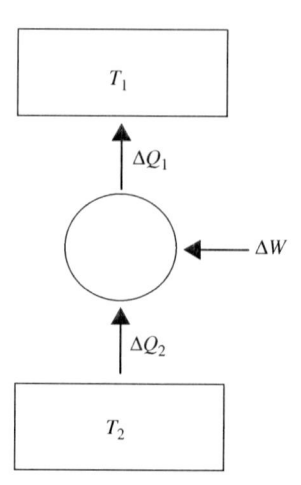

Fig. 6.1

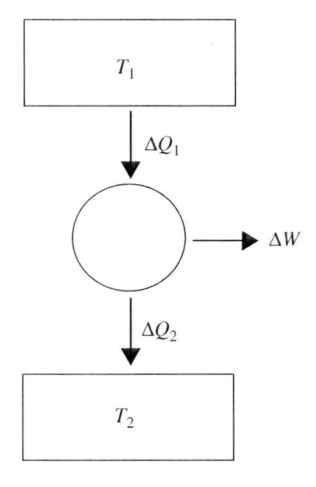

Fig. 6.2

It was Carnot who realized that the most efficient heat engine of all was a 'reversible' heat engine. In other words, one that got the same amount of work out of the heat transfer as would be needed to operate a perfect fridge to undo its operation.

In order to do this, it is necessary for all the heat transfers (between one object and another) to take place with as small a temperature difference as possible. If this is not done, heat will flow from hot objects to cold – a process which could have been used to do work, but was not. Therefore not enough work will be done to enable the fridge to return the heat to the hot object.

Carnot therefore proposed that the ratio of heat coming in from the hot object to the heat going out into the cold object has a maximum for this most efficient engine. This is because the difference between heat in and heat out is the work done, and we want to do as much work as possible.

So how would a 'Carnot' engine work? Whatever processes occur, there can never be any transfer of heat across a temperature difference (as this is irreversible). So let us have a go at designing such an engine. First, we can confine ourselves to two kinds of processes that were listed in Section 6.1:

(a) Isothermal processes (our heat engine can be attached to one of the reservoirs at the temperature of that reservoir and acquire ΔQ without changing temperature)
(b) Adiabatic processes (our heat engine can be isolated from the reservoirs and under an adiabatic change modify its temperature without involving an exchange of energy with the environment).

These processes are reversible because they do not involve the transfer of heat across a temperature difference.

A 'Carnot cycle' might look like the one shown in Figure 6.3.

120 *Thermal physics*

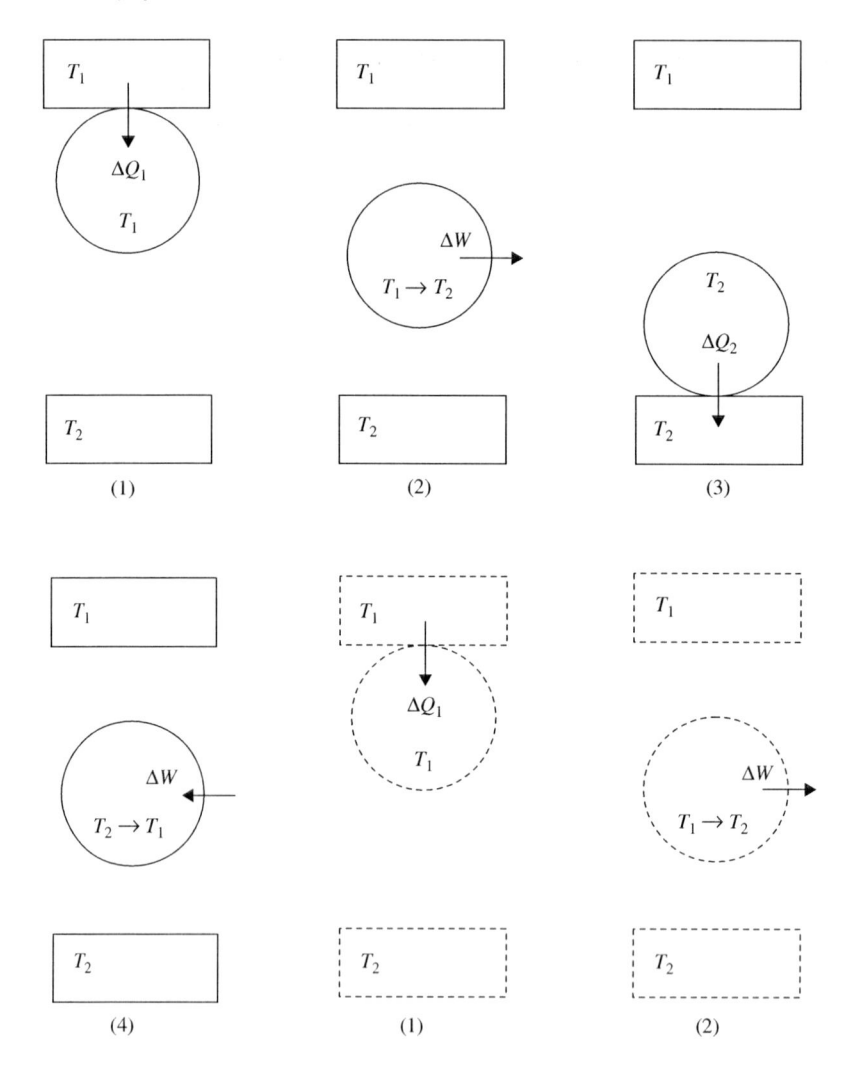

Fig. 6.3

(1) ΔQ_1 is transferred isothermally from the hotter reservoir to the engine at temperature T_1.
(2) The engine is isolated from the hotter reservoir and allowed to expand adiabatically. Work is done on the surroundings, but conservation of energy would require that the temperature of the engine must now fall. It falls from T_1 (the temperature of the hotter reservoir) to T_2 (the temperature of the colder reservoir).
(3) The engine is now attached to the colder reservoir and ΔQ_2 is transferred isothermally to the colder reservoir at temperature T_2.
(4) The engine is isolated from the colder reservoir and some work is done on it adiabatically. Since there is no transfer of heat (it is after all, adiabatic), the conservation of energy would require that the temperature of the engine will

rise from T_2 (the temperature of the colder reservoir) to T_1 (the temperature of the hotter reservoir).

After (4) we of course return to (1) and the beginning of the next cycle.

In thinking up this cycle, Carnot also noticed that processes (1) and (3) are essentially identical except in the temperatures at which the energies (ΔQ_1 and ΔQ_2) are transferred. Therefore, ΔQ_1 and ΔQ_2 can only be dependent on T_1 and T_2, respectively. Furthermore, he said that the ratio of the two energies must be a function of the temperatures of the hot and cold reservoirs only.

This can be stated as

$$\left|\frac{\Delta Q_1}{\Delta Q_2}\right| = f(T_1, T_2). \tag{6.5}$$

More light can be shed on the problem if we stack two heat engines in series, with the second taking the heat ΔQ_2 from the first (at temperature T_2), extracting further work from it before dumping it as heat (ΔQ_3) into a yet colder reservoir at temperature T_3.

The two heat engines separately and together give us the equations:

$$\left|\frac{\Delta Q_1}{\Delta Q_2}\right| = f(T_1, T_2) \quad \left|\frac{\Delta Q_2}{\Delta Q_3}\right| = f(T_2, T_3) \quad \left|\frac{\Delta Q_1}{\Delta Q_3}\right| = f(T_1, T_3)$$

$$\Rightarrow f(T_1, T_3) = f(T_1, T_2) \times f(T_2, T_3)$$

$$\Rightarrow f(T_1, T_2) = \frac{f(T_1, T_3)}{f(T_2, T_3)} = \frac{g(T_1)}{g(T_2)}. \tag{6.6}$$

6.3.2 Thermodynamic temperature

However, if $g(T)$ is a function of the temperature alone, we might as well call $g(T)$ the temperature itself. To summarize, thermodynamic temperature (T) is defined so that in a reversible heat engine, the ratio of heat extracted from the hot object (ΔQ_1) to the heat ejected into the cold object (ΔQ_2) is

$$\left|\frac{\Delta Q_1}{\Delta Q_2}\right| = \frac{T_1}{T_2}. \tag{6.7}$$

William Thomson[*] (Lord Kelvin) proposed a temperature scale based on this and Carnot's theory. Using the observations of a contemporary, Regnault, Thomson showed that for a close approximation to a Carnot cycle operating between the temperatures of boiling water (T_1) and pure melting ice (T_2) at atmospheric pressure,

$$\left|\frac{\Delta Q_1}{\Delta Q_2}\right| = 1.366.$$

With $T_1 - T_2 = 100$ units (Thomson wanted to match the Celsius scale), this is only achievable if $T_2 = 273.2$ units. This is of course the Kelvin scale, though nowadays by international agreement the 'Kelvin' and its scale are defined by two points: absolute

---

[*]On an absolute thermometric scale founded on Carnot's theory of the motive power of heat, and calculated from Regnault's observations. By Lord Kelvin (William Thomson) 1848.

122 Thermal physics

zero and the temperature of the triple point of water.* So $T_2 = 273.16$ K (or $0.01°$C) by modern standards, and is the triple point of water.

The 'Kelvin' temperature scale obtained using the gas laws (a perfect gas) satisfies this definition as well. For this reason, the kelvin is frequently referred to as the unit of 'thermodynamic temperature'. Section 6.6.2.3 considers a Carnot engine with a perfect gas as its working medium.

6.3.3 Efficiency of a heat engine

The efficiency of a reversible heat engine can now be calculated. We define the efficiency (η) to be the ratio of the work done (the useful output) to ΔQ_1 (the total energy input). Therefore,

$$\eta = \left|\frac{\Delta W}{\Delta Q_1}\right| = \frac{\Delta Q_1 - |\Delta Q_2|}{\Delta Q_1} = 1 - \frac{T_2}{T_1}. \tag{6.8}$$

This, being the efficiency of a reversible engine, is the maximum efficiency that can be achieved. A real engine will fall short of this goal. Notice that for a coal-fired power station, in which T_1 (the temperature of the boiler) is frequently 840 K, and T_2 (the temperature of the stream outside) is 300 K, the maximum possible efficiency is

$$\eta = 1 - \frac{300}{840} = 64\%.$$

In practice the water leaves the turbogenerator at 530 K, and so the efficiency cannot go higher than

$$\eta = 1 - \frac{530}{840} = 37\%.$$

The design of modern large power stations is such that the actual efficiency is remarkably close to this value.

Q1 Calculate the maximum efficiency possible in a coal-fired power station, if the steam is heated to $700°$C and the river outside is at $7°$C.

Q2 Mechanical engineers have been keen to build jet engines which run at higher temperatures. This makes it very difficult and expensive to make the parts, given that the materials must be strong, even when they are almost at their melting point. Why are they making life hard for themselves?

---

*The triple point of water is the state in which pure water, pure ice, and pure water vapour can coexist in stable equilibrium.

6.4 Entropy

6.4.1 Reversible processes

Now we need to take a step backwards before we can go forwards. Look back at the definition of thermodynamic temperature in equation (6.7). It can be rearranged to state:

$$\text{Reversible}: \qquad \frac{\Delta Q_1}{T_1} = \frac{|\Delta Q_2|}{T_2} \Rightarrow \frac{\Delta Q_1}{T_1} + \frac{\Delta Q_2}{T_2} = 0. \qquad (6.9)$$

Remember that this is for the ideal situation of a reversible process – as in a perfect fridge or heat engine. Suppose, then, that we start with some gas at pressure p and volume V. Then we do something with it (squeeze it, heat it, let it expand, or anything reversible), and finally do some more things to it to bring it back to pressure p and volume V. The list of processes can be broken up into tiny stages, each of which saw some heat (ΔQ) entering or leaving the system, which was at a particular temperature T. The only difference between this situation and that in (6.9) is that there were only two stages in the process for the simpler case. The physics of (6.9) should still apply, no matter how many processes are involved. Therefore, provided all the actions are reversible we can write

$$\text{Reversible}: \qquad \sum_{\text{Complete cycle}} \frac{\Delta Q}{T} = 0 \Rightarrow \oint \frac{dQ}{T} = 0, \qquad (6.10)$$

where the circle on the integral implies that the final position (on a p,V graph) is the same as where the gas started.

Now suppose that there are two points on the (p,V) graph which are of interest to us, and we call them A and B. Let us go from A to B and then back again (using a different route), but only using reversible processes. We call the first route I, and the second route II. Equation (6.10) tells us

$$\oint \frac{dQ}{T} = \int_A^B \frac{dQ_\text{I}}{T} + \int_B^A \frac{dQ_\text{II}}{T} = 0$$

$$\text{Reversible}: \qquad \int_A^B \frac{dQ_\text{I}}{T} - \int_A^B \frac{dQ_\text{II}}{T} = 0$$

$$\int_A^B \frac{dQ_\text{I}}{T} = \int_A^B \frac{dQ_\text{II}}{T}. \qquad (6.11)$$

In other words the integral $\int dQ/T$ between the two points A and B is the same, no matter which reversible route is chosen. This is a very special property of a function – we label $\int dQ/T$ as a function of states A and B, and call it the change in entropy, ΔS_{AB}.

124 Thermal physics

This means that the current entropy of the gas, like pressure, volume, and temperature, is only a function of the state that the gas is in now – and does not depend on the preparation method.

6.4.2 Irreversible processes and the second law

We must stress that entropy is only given by $\int dQ/T$ when the integral is taken along reversible processes in which there is no wastage of heat. Heat is wasted when it is allowed to flow from a hot object to a cold one without doing any work on the journey. This would be irreversible, since you could only get the heat back into the hot object if you expended more energy on it.

Let us make an analogy. Reversible processes are like a world in which purchasing prices and selling prices are the same. If you started with £100, and spent it in various ways, you could sell the goods and end up with £100 cash at the end.

Irreversible processes are like the real world in that traders will want to sell you an apple for more than they bought it for. Otherwise they will not be able to make a profit. If you started with £100, and spent it, you would never be able to get £100 back again, since you would lose money in each transaction. You may end up with £100 'worth of goods', but you would have to be satisfied with a price lower than £100 if you wanted to sell it all for cash.

Let us now return to the physics, and the gas in the piston. What does irreversibility mean here? We have not lost any energy – the first law has ensured that. But we have lost usefulness.

Equation (6.10) tells us that if we come back to where we started, and only use reversible processes on the way, the total entropy change will be zero. There is another way of looking at this, from the point of view of a heat engine.

Let us suppose that the temperature of the boiler in a steam engine is T_A. In a perfect heat engine, the cylinder will receive the steam at this temperature. Suppose ΔQ joules of heat are transferred from boiler to cylinder. The boiler loses entropy $\Delta Q/T_A$; the cylinder gains entropy $\Delta Q/T_A$ and the total entropy remains constant.

Now let us look at a real engine. The boiler must be hotter than the cylinder, or heat would not flow from boiler to cylinder. Suppose that the boiler is still at T_A, but the cylinder is at T_C. We have now let irreversibility loose in the system, since the heat ΔQ now flows from hot to cooler without doing work on the way.

What about the entropy? The boiler now loses $\Delta Q/T_A$ to the connecting pipe,* but the cylinder gains $\Delta Q/T_C$ from it. Since $T_C < T_A$, the cylinder gains more entropy than the boiler lost.

This is an alternative definition of the second law. Processes go in the direction to maximize the total amount of entropy.

---

*What has the pipe got to do with it? Remember that we said that change in entropy ΔS is only given by $\Delta Q/T$ for reversible processes. The passing of ΔQ joules of heat into the pipe is done reversibly (at temperature T_A), so we can calculate the entropy change. Similarly, the passing of ΔQ joules of heat from pipe to cylinder is done reversibly (at T_C), so the calculation is similarly valid at the other end. However, something is going on in the pipe which is not reversible – namely ΔQ joules of heat passing from higher to lower temperature. Therefore we cannot apply any $\Delta S = \Delta Q/T$ arguments inside the pipe.

6.4.3 Restatement of first law

For reversible processes, $\Delta W = -p\Delta V$, and $\Delta Q = T\Delta S$. Therefore the first law (6.1) can be written as

$$\Delta U = T\Delta S - p\Delta V. \tag{6.12}$$

We find that this equation is also true for irreversible processes. This is because T, S, U, p, and V are all functions of state, and therefore if the equation is true for reversible processes, it is true for all processes. However, care must be taken when using it for irreversible processes, since $T\Delta S$ is no longer equal to the heat flow and $p\Delta V$ is no longer equal to the work done.

Q3 Two insulated blocks of steel are identical except that one is at 0°C, while the other is at 100°C. They are brought into thermal contact. A long time later, they are both at the same temperature. Calculate the final temperature; the energy change and entropy change of each block if (1) heat flows by conduction from one block to the other, and if (2) heat flows from one to the other via a reversible heat engine. ++

6.5 The Boltzmann law

The Boltzmann law is simple to state, but profound in its implications:

$$\text{Probability that a particle has energy } E \propto e^{-E/kT}, \tag{6.13}$$

where k is the Boltzmann constant and is about 1.38×10^{-23} J/K. We also find that the probability that a system has energy E *or greater* is also proportional to $e^{-E/kT}$ (with a different constant of proportionality). This energy E is therefore very important and as such is often called an *activation energy* and processes that seem to obey this law are often called *activation processes*. If you are less familiar with the number e you might like to try the workshop in Section 7.2, which is on logarithms.

There is common sense here because (6.13) is saying that greater energies are less likely, and also that the higher the temperature, the more likely you are to have higher energies. The quotient in the power of the number e essentially compares the energy E with the energy kT. As we will see, typically, the average energy of a molecule in a system is given by kT, so (6.13) is very small if E/kT is large; that is, if T is small and kT is small compared to E – the likelihood of a molecule at low T having an energy E is small because the chances of a molecule receiving the energy from thermal excitations at low T is small. As T gets larger without limit we see that (6.13) tends to unity – now the average energy kT becomes much larger than the energy E and it becomes very likely that any molecule can have any energy.

Let us look at two examples.

6.5.1 Workshop: Atmospheric pressure

The pressure in the atmosphere at height h is proportional to the probability that a molecule will be at that height, and is therefore proportional to $e^{-mgh/kT}$. Here, the energy E is of course the gravitational potential energy of the molecule – which has mass m.

126 Thermal physics

The proof of this statement is in several parts.

First we assume that all the air is at the same temperature. This is not a realistic assumption, but we shall make do with it. Next we divide the atmosphere into slabs (each of height Δh and unit area), stacked one on top of the other. Each slab has to support all the ones above it.

(a) From the gas law* ($pV = NkT$ where N is the number of molecules under consideration), the definition of density ($\rho = Nm/V$) show that

$$\rho = \frac{pm}{kT}.$$

(b) Increasing the height h by a small increment Δh, the pressure will reduce by the weight of one slab. Show that

$$\Delta p = -\frac{pmg}{kT}\Delta h.$$

(c) Integrate the expression above to show that

$$p = p_0 e^{-(mgh/kT)} = p_0 e^{-(E_{\text{grav}}/kT)},$$

where p_0 is the pressure at ground level and E_{grav} is the difference in gravitational potential energy of a molecule at a height h and a molecule on the ground.

Sections 7.1 and 7.2 are workshops that deal with setting up integrals in physics problems and logarithms, respectively. You might like to have a go at them before you try (c).

We see that the Boltzmann law is obeyed for an isothermal atmosphere.

6.5.2 Velocity distribution of molecules in a gas

The probability that a molecule in the air will have an x-component of its velocity in a range u_x and $u_x + \Delta u_x$ is once again proportional to $e^{-E/kT}$. Here the energy E is the kinetic energy associated with the x-component of motion, namely $mu_x^2/2$. This means that the number of molecules with an x-component of velocity in the range u_x and $u_x + \Delta u_x$ is

$$\Delta N \propto e^{-mu_x^2/2kT}\Delta u_x. \qquad (6.14)$$

From this statement, we can set up the average value of u_x^2. Thus:

$$\overline{u_x^2} = \frac{\sum u_x^2 \Delta N}{\sum \Delta N}. \qquad (6.15)$$

The numerator of this quotient is just the sum of all the possible u_x^2 and the denominator is just the total number of molecules. Taking the limit as the interval Δu_x tends to zero we get an integral:

$$\overline{u_x^2} = \frac{\int u_x^2 \exp\left(-mu_x^2/2kT\right) du_x}{\int \exp\left(-mu_x^2/2kT\right) du_x} = \frac{\frac{1}{2}(2kT/m)^{3/2}}{(2kT/m)^{1/2}} = \frac{kT}{m}. \qquad (6.16)$$

*Dealt with in more detail in Section 6.6.

The mean kinetic energy is given by

$$\overline{K} = \frac{1}{2}m\overline{u^2} = \frac{1}{2}m\overline{(u_x^2 + u_y^2 + u_z^2)} = \frac{3}{2}kT, \qquad (6.17)$$

as (6.16) would have produced the same answer for any direction (x, y, or z) in space. So the internal energy of a mole of gas (due to linear motion) is

$$U = N_A \overline{K} = \frac{3}{2} N_A kT = \frac{3}{2} RT, \qquad (6.18)$$

where $R \equiv N_A k$ is the gas constant. From this it follows that the molar heat capacity of a perfect gas,* $C_V = \frac{3}{2}R$.

Q4 There is a 'rule of thumb' in chemistry that when you raise the temperature by 10°C, the rate of reaction roughly doubles. Use Boltzmann's law to show that this means the activation energy of chemical processes must be of order 10^{-19} J. +

Q5 The amount of energy taken to turn 1 kg of liquid water at 100°C into 1 kg of steam at the same temperature is 2.26 MJ. This is called the latent heat of vaporization of water. How much energy does each molecule need to 'free itself' from the liquid?

Q6 The probability that a water molecule in a mug of tea has enough (or more than enough) energy to leave the liquid is proportional to $\exp(-E_L/kT)$ where E_L is the energy required to escape the attractive pull of the other molecules (latent heat of vaporization per molecule). By definition, the boiling point of a liquid is the temperature at which the saturated vapour pressure is equal to atmospheric pressure (about 100 kPa). Up a mountain, you find that you cannot make good tea, because the water is boiling at 85°C. What is the pressure? You will need your answer to Q5. +

Q7 Estimate the altitude of the mountaineer in Q6. Assume that all of the air in the atmosphere is at 0°C. +

Q8 The fraction of molecules (mass m each) in a gas at temperature T which have a particular velocity (of speed u) is proportional to $e^{-mu^2/2kT}$, as predicted by the Boltzmann law. However, the fraction of molecules which have speed u is proportional to $u^2 e^{-mu^2/2kT}$. Where does the u^2 come from? ++

6.5.3 Workshop: Justification of Boltzmann law

In this section, we aim to give you an understanding about where the $e^{-E/kT}$ comes from in fundamental terms. This is hardly necessary in most first-year university courses; however, we feel you deserve some kind of explanation. As a benefit, this analysis also introduces you to some of the most vital concepts in statistical mechanics.

Imagine you toss three fair coins. There are two ways of describing the outcome. You could simply say, 'Two heads, one tail', or you could be more specific: 'coins

---

*This is the heat capacity due to linear motion. For a monatomic gas (like helium), this is the whole story. For other gases, the molecules can rotate or vibrate about their bonds as well, and therefore the heat capacity will be higher.

A and B came down heads, coin C came down tails'. The first description is called a macrostate and the second a microstate. Notice that if the coins were identical, you could not distinguish the two 'head' coins from each other, and therefore the microstate A = heads, B = tails, C = heads would look the same as A = B = heads, C = tails.

(a) Write down all the possible macrostates for three tossed coins. Work out how many microstates each one contains.
(b) Now suppose instead of tossed coins, we have 4 units of energy to give to seven atoms. The units cannot be split up, but we can give one atom more than one unit of energy if we wish. Two of the resulting macrostates are 'one atom with three energy units, one atom with one unit, five atoms with no energy', and 'four atoms with one energy unit, three with none'. Write down the other three possible macrostates.
(c) A macrostate can be more conveniently written $\{n_0, n_1, n_2, n_3, \ldots\}$ where n_0 is the number of atoms with no units of energy, n_1 gives the number of atoms with only one unit, and n_p gives the number of atoms with exactly p units of energy. Accordingly, the two macrostates written out above could be rewritten $\{5,1,0,1\}$ and $\{3,4\}$, respectively. Use this notation to write down the other three macrostates.
(d) Attempt to calculate the number of microstates in each macrostate.*
(e) If all microstates were equally likely, which macrostate would be most likely? What do you notice about this macrostate?

It turns out that the most likely macrostate will always be the one in which the numbers n_0, n_1, n_2, ... are closest to a geometric progression. In other words, for the most likely macrostate, $n_p = Af^p$, where f is some fixed number.† Given that there are fewer atoms with larger amounts of energy, the number f will be less than 1. Often this relationship is written $n_p = Ae^{-\alpha p}$ where α is a positive constant.

(f) Derive a relationship between α and f in the previous paragraph.
(g) If the number of atoms with p units of energy is given by $n_p = Af^p$ as above, prove that the total number of atoms $N = \sum_p n_p = A/(1-f)$.‡
(h) If the number of atoms with p units of energy is given by $n_p = Af^p$ as above, prove that the total number of energy units $E = \sum_p p n_p = Af/(1-f)^2$.§
(i) Using your answers to (g) and (h), derive expressions for A and f in terms of N and E. Hence write down the value of n_p in terms of N and ε where $\varepsilon = E/N$

---

*The formula for the number of microstates in the macrostate is given by $W = N!/(n_0!n_1!n_2!\ldots)$ where the exclamation mark means 'factorial' – i.e. $3! = 3 \times 2 \times 1$, $4! = 4 \times 3 \times 2 \times 1$, and so on (0! is defined equal to 1); and N is the total number of atoms. You might like to think about why this might be the case. A justification is given in the solution to part (d) of this workshop.
†This is easier to prove than you might think, and a justification is given in the solution to part (e) of this workshop.
‡Hint: Note that $Nf = N - A$.
§Hint: Note that $(f + 2f^2 + 3f^3 + 4f^4 + \cdots) = (1 + f + f^2 + f^3 + \cdots)S$ where $S = f + f^2 + f^3 + f^4 + \cdots$, and then use your answer to the previous question to sum the geometric series.

is the average number of energy units per atom. You should find that

$$n_p = N\left(\frac{1}{1+\varepsilon}\right) \times \left(\frac{\varepsilon}{1+\varepsilon}\right)^p = \frac{N}{1+\varepsilon} \times \left(1+\varepsilon^{-1}\right)^{-p}.$$

(j) When we use this method to describe the distribution of energy between many atoms, we frequently make an assumption that E is much larger than N, so that the amount of energy which an atom has can be regarded as continuous. Accordingly ε is much larger than 1. Show that in this case, the α factor of question (f) becomes $\alpha = \varepsilon^{-1}$, and hence:

$$n_p = Ae^{-p/\varepsilon}.$$

(k) Now suppose that each unit of energy is actually η joules worth. Thus if an atom has p units of energy, it has $p\eta$ joules (which we now define as F); while the average energy per atom is $E\eta/N = \varepsilon\eta$ (which we now define as $\overline{F}$). Rewrite the equation for n_p in terms of A, F, and $\overline{F}$. Typically, the average energy is given by kT, and hence we find that the Boltzmann distribution comes about because it is the most likely macrostate.

6.6 Perfect gases

All substances have an *equation of state*. This tells you the relationship between volume, pressure, and temperature for the substance. Most equations of state are nasty; however, the one for an ideal, or perfect, gas is straightforward to use. It is called the gas law. This states that

$$pV = nRT \qquad (6.19)$$
$$pV = NkT, \qquad (6.20)$$

where p is the pressure of the gas, V its volume, and T its absolute (or thermodynamic) temperature. This temperature is measured in kelvin *always*. There are two ways of stating the equation: as in (6.19), where n represents the number of moles of gas, or as in (6.20), where N represents the number of molecules of gas. Clearly $N = N_A n$ where N_A is the Avogadro number, and therefore $R = N_A k$.

You can adjust the equation to give you a value for the number density of molecules. This means the number of molecules per cubic metre, and is given by $N/V = p/kT$. The volume of one mole of molecules can also be worked out by setting $n = 1$ in (6.19):

$$V_m = \frac{RT}{p}. \qquad (6.21)$$

You can adjust this equation to give you an expression for the density. If the mass of one molecule is m, and the mass of a mole of molecules (the relative molecular mass) is M, we have

$$\rho = \frac{\text{Mass}}{\text{Volume}} = \frac{M}{RT/p} = \frac{Mp}{RT}$$
$$= \frac{N_A mp}{N_A kT} = \frac{mp}{kT}. \tag{6.22}$$

Please note that this is the ideal gas law. Real gases will not always follow it. This is especially true at high pressures and low temperatures where the molecules themselves take up a good fraction of the space. However, at room temperature and atmospheric pressure, the gas law is a very good model.

Q9 Use the gas law to work out the volume of one mole of gas at room temperature and pressure (25°C, 100 kPa).

Q10 What fraction of the volume of the air in a room is taken up with the molecules themselves? Make an estimate, assuming that the molecules are about 10^{-10} m in radius.

Q11 Estimate a typical speed for a nitrogen molecule in nitrogen at room temperature and pressure. On average, how far do you expect it to travel before it hits another molecule? Again, assume that the radius of the molecule is about 10^{-10} m. ++

6.6.1 Heat capacity of a perfect gas

We have already shown (in Section 6.5.2) that for a perfect gas, the internal energy due to linear motion is $\frac{3}{2}RT$ per mole. If this were the only consideration, then the molar heat capacity would be $\frac{3}{2}R$. However, there are two complications which are described below.

6.6.1.1 The conditions of heating

In thermodynamics, you will see molar heat capacities written with subscripts: C_P and C_V. They both refer to the energy required to heat a mole of the substance (M kg) by 1 K. However, the energy needed is different depending on whether the volume or the pressure is kept constant as the heating progresses.

When you heat a gas at constant volume, all the heat going in goes into the internal energy of the gas ($\Delta Q_V = \Delta U$).

When you heat a gas at constant pressure, two things happen. The temperature goes up, but it also expands. In expanding, it does work on its surroundings. Therefore the heat put in is increasing both the internal energy and is also doing work ($\Delta Q_P = \Delta U + p\Delta V$).

Given that we know the equation of state for the gas (6.19), we can work out the relationship between the constant-pressure and constant-volume heat capacities. In these equations we shall be considering one mole of gas.

$$\Delta Q_V = C_V \Delta T = \frac{dQ_V}{dT}\Delta T = \frac{dU}{dT}\Delta T$$
$$\Delta Q_P = C_P \Delta T = \frac{dQ_P}{dT}\Delta T = \frac{dU}{dT}\Delta T + p\frac{dV}{dT}\Delta T$$

so

$$C_\text{V} = \frac{dQ_\text{V}}{dT} = \frac{dU}{dT}$$
$$C_\text{P} = \frac{dQ_\text{P}}{dT} = \frac{dU}{dT} + p\frac{dV}{dT}$$
$$= C_\text{V} + p\frac{d}{dT}\left(\frac{RT}{p}\right) = C_\text{V} + R. \quad (6.23)$$

6.6.1.2 The type of molecule

Gas molecules come in many shapes and sizes. Some only have one atom (like helium and argon), and these are called monatomic gases. Some gases are diatomic (like hydrogen, nitrogen, oxygen, and chlorine), and some have more than two atoms per molecule (like methane).

The monatomic molecule only has one use for energy – going places fast. Therefore a monatomic molecule's internal energy is given simply by $\frac{3}{2}kT$, and so the molar internal energy is $U = \frac{3}{2}RT$. Therefore, using equation (6.23), we can show that $C_\text{V} = \frac{3}{2}R$ and $C_\text{P} = C_\text{V} + R = \frac{5}{2}R$.

A diatomic molecule has other options open to it. The atoms can rotate about the molecular centre (and have a choice of two axes of rotation). They can also wiggle back and forth – stretching the molecular bond like a rubber band. At room temperature we find that the vibration does not have enough energy to kick in, so only the rotation and translation (the linear motion) affect the internal energy.

Each possible axis of rotation adds $\frac{1}{2}kT$ to the molecular energy, and so we find that for most diatomic molecules, $C_\text{V} = \frac{5}{2}R$ and $C_\text{P} = \frac{7}{2}R$.

6.6.1.3 Thermodynamic gamma

It turns out that the ratio of C_P/C_V crops up frequently in equations, and is given the letter γ. This is not to be confused with the γ in relativity, which is completely different.

Using the results of our last section, we see that $\gamma = 5/3$ for a monatomic gas, and $\gamma = 7/5$ for one that is diatomic.

6.6.2 Pumping heat

In this section we show you how to turn a perfect gas (in a cylinder) into a reversible heat engine, and in doing so we will summarize all that we have dealt with so far.

6.6.2.1 Isothermal gas processes

As an introduction, we need to know how to perform two processes. First we need to be able to get heat energy into or out of a gas without changing its temperature. Remember that we want a reversible heat engine, and therefore the gas must be at the same temperature as the hot reservoir when the heat is passing into it. Any process, like this, which takes place at a constant temperature is said to be *isothermal*.

132   *Thermal physics*

The gas law tells us (6.19) that $pV = nRT$, and hence that pV is a function of temperature alone (for a fixed amount of gas). Hence in an isothermal process,

$$pV = \text{constant}. \tag{6.24}$$

Using this equation, we can work out how much we need to compress the gas to remove a certain quantity of heat from it. Alternatively, we can work out how much we need to let the gas expand in order for it to 'absorb' a certain quantity of heat. These processes are known as isothermal compression and isothermal expansion, respectively.

Suppose that the volume is changed from V_1 to V_2, the temperature remaining T. Let us work out the amount of heat absorbed by the gas. First of all, remember that as the temperature is constant, the internal energy will be constant, and therefore the first law may be stated $\Delta Q = p\Delta V$. In other words, the total heat entering the gas may be calculated by integrating $p\Delta V$ from V_1 to V_2:

$$Q = \int p\,dV = \int \frac{nRT}{V}\,dV = [nRT \ln V]_{V_1}^{V_2} = nRT \ln \frac{V_2}{V_1}. \tag{6.25}$$

This equation describes an isothermal (constant temperature) process only. In order to keep the temperature constant, we maintain a good thermal contact between the cylinder of gas and the hot object (e.g. the boiler wall) while the expansion is going on.

6.6.2.2   Adiabatic gas processes

The other type of process you need to know about is the adiabatic process. These are processes in which there is no heat flow ($\Delta Q = 0$), and they are used in our heat engine to change the temperature of the gas in between its contact with the hot object and the cold object. Sometimes this is referred to as an isentropic process, since if $\Delta Q = 0$ for a reversible process, $T\Delta S = 0$, and so $\Delta S = 0$ and the entropy remains unchanged.[*]

Before we can work out how much expansion causes a certain temperature change, we need to find a formula which describes how pressure and volume are related in an adiabatic process. First, the first law tells us that if $\Delta Q = 0$, then $0 = \Delta U + p\Delta V$. We can therefore reason like this for n moles of gas:

$$0 = \Delta U + p\Delta V$$
$$= n\,C_V\,\Delta T + p\,\Delta V.$$

---

[*]While the terms 'isentropic' and 'adiabatic' are synonymous for a perfect gas, care must be taken when dealing with irreversible processes in more advanced systems. In this context ΔQ is not equal to $T\Delta S$. If $\Delta Q = 0$, the process is said to be adiabatic; if $\Delta S = 0$, the process is isentropic. Clearly for a complex system, the two conditions will be different. This arises because in these systems, the internal energy is not just a function of temperature, but also of volume or pressure.

Now for a perfect gas, $nRT = pV$, therefore $nR\Delta T = p\Delta V + V\Delta p$. So we may continue the derivation thus:

$$\begin{aligned} nC_V\Delta T &= \frac{C_V}{R}(p\Delta V + V\Delta p) \\ 0 &= \frac{C_V}{R}(p\Delta V + V\Delta p) + p\Delta V \\ &= C_V(p\Delta V + V\Delta p) + Rp\Delta V \\ &= C_P p\Delta V + C_V V\Delta p \\ &= \gamma p\Delta V + V\Delta p \\ &= \gamma \frac{\Delta V}{V} + \frac{\Delta p}{p}. \end{aligned} \qquad (6.26)$$

Integrating* this differential equation gives

$$\gamma \ln V + \ln p + C = 0$$
$$pV^\gamma = e^{-C}$$
$$pV^\gamma = \text{constant}. \qquad (6.27)$$

Equation (6.27) is our most important equation for adiabatic gas processes, in that it tells us how pressure and volume will be related during a change.

We now come back to our original question: what volume change is needed to obtain a certain temperature change? Let us suppose we have a fixed amount of gas (n moles), whose volume changes from V_1 to V_2. At the same time, the temperature changes from T_1 to T_2. We may combine equation (6.27) with the gas law to obtain

$$pV^\gamma = \text{constant}$$
$$pV\,V^{\gamma-1} = \text{constant}$$
$$nRTV^{\gamma-1} = \text{constant} \qquad (6.28)$$
$$\frac{T_1}{T_2}\frac{V_1^{\gamma-1}}{V_2^{\gamma-1}} = 1.$$

6.6.2.3 A gas heat engine

We may now put our isothermal and adiabatic processes together to make the heat engine, which we first introduced in Section 6.3.1. Remember, the engine operates on a cycle:

(1) The cylinder is attached to the hot reservoir (temperature T_1), and isothermal expansion is allowed (from V_1 to V_2) so that heat ΔQ_1 is absorbed into the gas.

---

*This step involves both integration and natural logarithms. If you are less than familiar with either of these concepts, please have a go at the workshops in Sections 7.1 and 7.2.

134 *Thermal physics*

(2) The cylinder is detached from the hot reservoir, and an adiabatic expansion (from V_2 to V_3) is allowed to lower the temperature to that of the cold reservoir (T_2).
(3) The cylinder is then attached to the cold reservoir. Heat ΔQ_2 is then expelled from the cylinder by an isothermal compression from V_3 to V_4.
(4) Finally, the cylinder is detached from the cold reservoir. An adiabatic compression brings the volume back to V_1, and the temperature back to T_1.

Applying equation (6.25) to the isothermal processes gives us

$$\Delta Q_1 = nRT_1 \ln \frac{V_2}{V_1}$$
$$\Delta Q_2 = nRT_2 \ln \frac{V_4}{V_3}. \tag{6.29}$$

Similarly, applying equation (6.28) to the adiabatic processes gives us

$$\frac{T_1}{T_2} = \left(\frac{V_3}{V_2}\right)^{\gamma-1}$$
$$\frac{T_1}{T_2} = \left(\frac{V_4}{V_1}\right)^{\gamma-1}$$
$$\Rightarrow \frac{V_3}{V_2} = \frac{V_4}{V_1} \Rightarrow \frac{V_3}{V_4} = \frac{V_2}{V_1}. \tag{6.30}$$

Notice that this result means that the ratio of ΔQ_1 and ΔQ_2, depends *only* on the temperatures of the heat reservoirs T_1 and T_2. Combining equations (6.29) and (6.30) gives us

$$\frac{\Delta Q_1}{\Delta Q_2} = -\frac{T_1}{T_2}$$
$$\left|\frac{\Delta Q_1}{\Delta Q_2}\right| = \frac{T_1}{T_2}, \tag{6.31}$$

where the minus sign reminds us that $Q_2 < 0$, since this heat was leaving the gas.

To summarize this process, we have used a perfect gas to move heat from a hot reservoir to a colder one. In doing this, we notice less heat was deposited in the cold reservoir than absorbed from the hot one. Where has it gone? It materialized as useful work when the cylinder was allowed to expand. Had the piston been connected to a flywheel and generator, we would have seen this in a more concrete way.

We also notice that we have proved that the Kelvin scale of temperature, as defined by the gas law, is a true thermodynamic temperature since equation (6.31) is identical to (6.7).

Q12 A volume V of gas is suddenly squeezed to one-hundredth of its volume. Assuming that the squeezing was done adiabatically, calculate the work done on the gas, and the temperature rise of the gas. Why is the adiabatic assumption a good one for rapid processes such as this?

6.7 Conclusion

We began this chapter with the first law of thermodynamics, which is essentially a statement of the law conservation of energy. We make the distinction between the two laws because the first law is explicit about the relationship between work, internal energy, and heat.

Heat will flow from a hotter object to a colder object and will never, by itself, do the reverse. A fridge will do the reverse, but only if it is plugged in; that is, only if you are prepared to do some work will you be able to transfer heat from the icebox (colder object) to the surroundings (hotter object). This is essentially the second law of thermodynamics.

A heat engine allows heat to flow through it and arranges to do some work on the way. If the processes involved in the operation of the heat engine are all reversible, then the heat engine is the most efficient it can be. Irreversible processes involve heat flowing from hot objects to cold – a process which could have been used to do work, but was not. The *Carnot engine* is this idealized reversible engine and close approximations of the *Carnot cycle* may be used to define the Kelvin temperature scale. When we devise a Carnot engine with a perfect gas as the working medium, we discover that the Kelvin scale as defined by the ideal gas law is identical to the scale defined by a general Carnot cycle. The Carnot cycle and the idea of reversibility allow us to introduce the concept of entropy, thereby quantifying ideas introduced earlier in the chapter.

The Boltzmann law gives insight into many physical phenomena. Very often such phenomena are called *activation phenomena* as the probability of occurrence becomes significant when the available thermal energy becomes a good fraction of an *activation energy*. We also presented a justification of the Boltzmann law in the form of a workshop.

7
Miscellany

7.1 Workshop: Setting up integrals

Problems that require integration in physics are often difficult for beginners *not* because the integrals are so difficult (in fact the integrals themselves are usually quite easy), but because it takes a *lot* of practice to become comfortable with the process of translating the *physical* problem statement into the corresponding *mathematical* integral. This is called 'setting up' the integral. Let us use an example to illustrate the idea.

Figure 7.1 shows a dam holding back water of depth h. The water exerts a horizontal force on the dam but this force is larger at the base of the dam than near the surface of the water. Neglecting atmospheric pressure (as it acts on either side of the dam in the same way), we should be able to calculate the resultant horizontal force due to the water by summing up the contributions from the water at different depths.

The thing to do is to imagine the force exerted by the water on infinitesimal elements of area $l\Delta y$ as shown in Figure 7.1, and to 'add them up' (i.e. to integrate) to get the total effect.

We need a clear diagram (like Figure 7.1), showing a representative example of one of the infinitesimal elements and labelling it and any auxiliary coordinate (y) that might be helpful in specifying the location of this or any other element. It is critical that one does not choose an element that is *special* in any way (e.g. the element at $y = h$). That way the formula one obtains is applicable to any other element. The infinitesimal horizontal force ΔF on one of these elements is

$$\Delta F = \text{Pressure} \times \text{area} = \rho g y \times l \Delta y.$$

Now,

$$\Delta F = \frac{dF}{dy} \Delta y$$

is an approximate expression for ΔF for small Δy and the approximation gets better as $\Delta y \to 0$, so in this limit

$$\frac{dF}{dy} = \rho g l y.$$

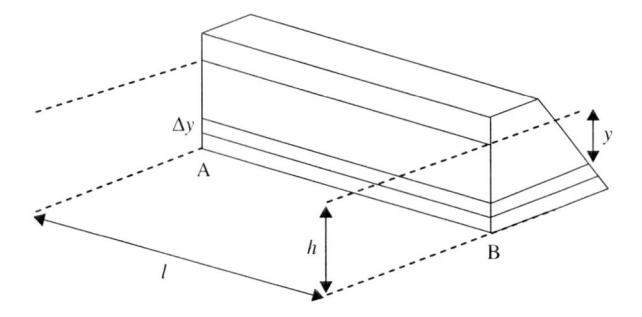

Fig. 7.1

So, the total resultant force F is the limiting sum

$$F = \int_0^h \frac{dF}{dy} dy = \rho g l \int_0^h y\, dy = \frac{\rho g l h^2}{2}.$$

(a) Obtain an expression for the resultant moment about AB caused by the water on the dam. (*Hint*: The perpendicular distance of an element of area from AB is $(h-y)$.)

(b) Calculate the height at which the resultant force would have to act to produce the same total moment.

Occasionally one is also required to construct integrals over multiple variables. A volume integral is a typical example of this sort of problem.

An element of volume of the sphere is depicted in Figure 7.2 as a prism made of sides Δr, $r\Delta \theta$, and $r \sin \theta \Delta \varphi$. In this coordinate system, called *spherical polar*, the

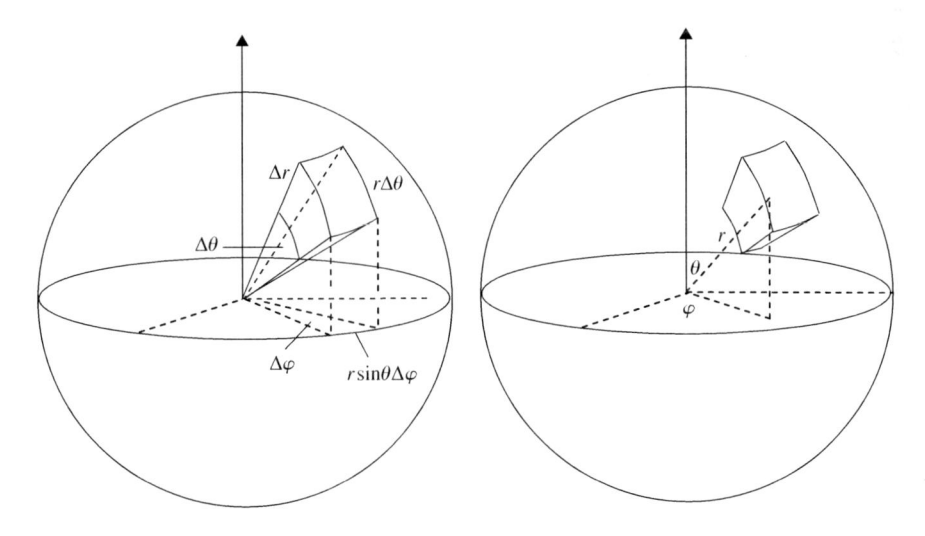

Fig. 7.2

138 Miscellany

element is referred to by the coordinates (r, θ, φ). Hence the volume V of the whole sphere is calculated by adding up all the volume elements

$$r^2 \sin\theta \, \Delta r \, \Delta\theta \, \Delta\varphi$$

and taking the limit as $\Delta r \to 0$, $\Delta\theta \to 0$, and $\Delta\varphi \to 0$ we have

$$V = \iiint r^2 \sin\theta \, dr \, d\theta \, d\varphi.$$

Since the variables are independent of each other, the three integrals can be performed separately and V becomes the product of the three integrals:

$$V = \int_0^R r^2 \, dr \int_0^\pi \sin\theta \, d\theta \int_0^{2\pi} d\varphi.$$

(c) Show that this gives the well-known expression: $V = \frac{4}{3}\pi R^3$.

7.2 Workshop: Logarithms

The common logarithm (log) of a number is more clearly demonstrated than defined. Look at the table below:

Number (x)	Can be written as	Common logarithm ($\log x$)
100	10^2	2
1000	10^3	3
0.1	10^{-1}	-1
3.162...	$10^{1/2}$	0.5
2	$10^{0.301...}$	0.301...

The log of a number is the power to which 10 would have to be raised to make that number. Thus $y = \log(x)$ is the solution to the equation $x = 10^y$.

(a) What is the log of 100 000? What is the log of 10 000? What is the log of 10 000 × 100 000? What do you notice?

(b) What is the log of 100 000 ÷ 10 000. What do you notice?

The questions should have convinced you of the truth of the following three statements:

$$\log(xy) = \log x + \log y, \tag{7.1}$$
$$\log(x/y) = \log x - \log y, \tag{7.2}$$
$$\log(x^a) = a \log x. \tag{7.3}$$

The first follows because $10^{x+y} = 10^x \times 10^y$, and the other two follow by extension.

(c) If $\log 2 = 0.301$ and $\log 3 = 0.477$ to three decimal places, work out the logs of the following numbers without a calculator, then check: (a) 6, (b) 1.5, (c) 4, (d) 9, and (e) 0.5.

(d) Sketch (or plot) the graph of $\log y$ against $\log x$ where $y = 100x^4$. Why do you get a straight line? What are the values of the y-intercept and gradient? Why?

(e) If I start with mass m_0 of a radioisotope with half-life T, I have only $m_0/2$ remaining after one half-life and $m_0/4$ after two half-lives. Show that the mass of radioisotope remaining after time t is equal to

$$m = m_0 \left(\frac{1}{2}\right)^{t/T}. \qquad (7.4)$$

Show that if I plot the graph of $\log m$ against t, I get a straight line. The y-intercept is related to m_0 and the gradient related to T. Find the relationships.

You may be wondering what was special about the number 10 (called the *base* of the logarithms). Why did we say that $\log x = y$ meant that $10^y = x$ rather than $2^y = x$ or $2.7^y = x$? The answer is that we generally use a denary counting system (base 10) and therefore this choice makes our logarithms as easy as possible for the common man or woman. Therefore they are called common logarithms, and not too long ago, before the age of the calculator, everyone used them when they worked out what a 7.5% reduction in the price of some trousers meant. These days logarithms are not so common, and we need not just use a base of 10.

We can use any number as a base, but when doing so, we need to specify the base. Thus since $1024 = 2^{10}$, we say that $\log_2 1024 = 10$, with the little 2 specifying the base. If no base is given, we assume that common logarithms (base 10) are in use.

(f) What is the logarithm of 56 in base 2? *Hint*: This means that $56 = 2^x$. We can use equation (7.3) to help us find x.

There is one base that is far more important mathematically than all of the others put together. It turns out that logarithms to base $e = 2.71\ldots$ are particularly useful in mathematics, and these are called natural logarithms. The natural logarithm of a number is usually written ln (standing for log natural). For example, $\ln 10 = 2.30$ to three significant figures.

$$\text{If } y = e^x, \text{ then } x = \ln y. \qquad (7.5)$$

The importance of these particular logarithms stems from the fact that the natural logarithm of x is the solution of the following integral:

$$\ln x = \int_{y=1}^{x} \frac{1}{y} dy. \qquad (7.6)$$

(g) Show that $\ln(ab) = \ln a + \ln b$ using the integral in equation (7.6). This proves that the integral of $1/y$ has the properties of a logarithm stated in equation (7.1).*

Hint: Start with $\ln ab = \int_1^{ab} \frac{1}{y} dy = \int_1^{a} \frac{1}{y} dy + \int_a^{ab} \frac{1}{y} dy = \ln a + \int_a^{ab} \frac{1}{y} dy$ and then substitute $y = za$ in the remaining integral, changing the dummy variable from y to z.

140 Miscellany

It follows from equation (7.6) that

$$\frac{d}{dx}\ln x = \frac{1}{x}. \tag{7.7}$$

If we write $x = e^y$, where e is defined to be the base of the natural logarithms; it follows that $\ln x = y$, and accordingly,

$$\frac{dy}{dx} = \frac{1}{e^y} \Rightarrow \frac{dx}{dy} = e^y \Rightarrow \frac{d}{dy}e^y = e^y. \tag{7.8}$$

Thus the function e^x is a function which, when differentiated with respect to x remains unchanged.

(h) Show that any function of the form Ae^x remains unchanged on differentiation with respect to x, providing that A is a constant.

(i) Show that

$$\frac{d}{dx}Ae^{kx} = kAe^{kx},$$

and thus that a solution of the differential equation $dy/dx = ky$ is given by $y = Ae^{kx}$, where A is a constant to be determined from the situation.

(j) If we assume that the function e^x can be written in the form

$$e^x = a_0 + a_1 x + a_2 x^2 + a_3 x^3 + \cdots,$$

where the a_p are constants, then
  (i) explain why $a_0 = 1$
  (ii) show that it follows from equation (7.8) that $a_{p-1} = pa_p$,
  (iii) show that $a_p = 1/p!$, and thus that $e^x = 1 + x + \frac{1}{2}x^2 + \frac{1}{6}x^3 + \cdots$,
  (iv) finally, by substituting $x = 1$, show that $e = e^1 = 2.718\ldots$.

If you want more practice with logarithms, you may like to try solving workshop 4.11.3 using logs.

7.3 Workshop: Rockets and stages

In (1.67) we recovered the well-known vector equation $\mathbf{F} = m\mathbf{a}$ from the full derivative of the momentum vector $\mathbf{p} = m\mathbf{v}$. This was because m, in the simple case we were studying (the mass of a falling body), was constant so $(d/dt)m = 0$. However, how do we deal with the situation when $(d/dt)m \neq 0$? One situation in which this condition occurs is when we study the acceleration of a rocket. A rocket works by ejecting combustion gases out of the back end (hence the mass of the rocket changes), which drives the body of the rocket forwards. Here we are going to consider the performance of a rocket in force-free space; that is, far from any other gravitating body so that all external forces are assumed to be zero.

A rocket at an initial velocity of $\mathbf{v}$ and total mass $m + M$ ejects a small mass of gas m and acquires a new velocity $\mathbf{v} + \Delta\mathbf{v}$. The ejected mass of gas leaves the rocket at a velocity relative to the ground of $\mathbf{v} - \mathbf{v_e}$, where $\mathbf{v_e}$ is the velocity of ejection of the exhaust gas. With no external forces acting on the system we once again have

$$\mathbf{F} = \frac{d}{dt}\mathbf{P} = \mathbf{0},$$

where $\mathbf{P}$ is the total momentum of the system.

(a) Show that momentum conservation leads to the following relationship:

$$\mathbf{0} = -m\mathbf{v_e} + M\Delta\mathbf{v}.$$

The small mass of fuel m is effectively a small reduction in mass of the rocket $-\Delta M$. So*:

$$-\frac{\Delta M}{M} = \frac{\Delta v}{v_e}.$$

Here we have dropped the vector notation as we understand $\Delta\mathbf{v}$ and $\mathbf{v_e}$ to be vectors along the line of flight. This expression can be integrated to obtain an expression for the sum of all the small boosts in speed Δv from an initial speed v_0 to a final speed v_1:

$$-v_e \int_{M_0}^{M_1} \frac{dM}{M} = \int_{v_0}^{v_1} dv,$$

where M_0 and M_1 are, respectively, the initial and final masses of the rocket before and after the burning of a large amount of fuel.

(b) Perform the integral and show that[†]

$$\frac{v_1 - v_0}{v_e} = \ln\left(\frac{M_0}{M_1}\right).$$

The rocket expends all of its fuel to achieve a velocity increment $(v_1 - v_0)$. The mass of the rocket in the absence of fuel is made up of a structural mass M_S and a payload mass M_L.

Rocket engineers find it convenient to introduce dimensionless parameters:

(i) The 'dead weight fraction' or s is the ratio of the structural mass to the initial mass of the rocket before firing its rocket motor *without* its payload.
(ii) The 'payload fraction' or l is the ratio of the payload mass to the initial mass of the rocket before firing its motor *with* its payload.

(c) Show that the velocity increment may be rewritten in terms of s and l as

$$\frac{v_1 - v_0}{v_e} = -\ln(s(1-l) + l).$$

---

[*]You might like to have a go at Workshop 7.1, which is a workshop on setting up integrals in physics problems.
[†]You might like to take a look at Section 7.2, which is just to remind you of the properties of this function.

Miscellany

In order to improve the performance of the rocket, several rockets may be combined to give a multistage vehicle. After the first stage is fired and its fuel expended it is jettisoned and the second stage is ignited, and so on. The advantage of the multistage rocket is that the structural weight at any time is better scaled to the amount of propellant that remains.

To determine the performance of a multistage rocket, the performance of each individual stage is obtained using the basic equation for the velocity increment of a single-stage rocket. The initial mass of the nth stage, $M_0^{(n)}$, is made up of the propellant mass used by the nth stage, $M_P^{(n)}$, the structural mass of the nth stage, $M_S^{(n)}$, and the initial mass of the next stage, $M_0^{(n+1)}$,

$$M_0^{(n)} = M_P^{(n)} + M_S^{(n)} + M_0^{(n+1)}.$$

The notation introduced for the single-stage rocket may be extended to the multistage rocket if it is noted that the payload of the nth stage is the initial mass of the next stage:

$$M_L^{(n)} = M_0^{(n+1)}.$$

(d) By assuming that v_e (the velocity of ejection of exhaust gases), s (the dead weight fraction), and l (the payload fraction) are the same for all stages of an N-stage rocket, show that the total velocity increment after N stages is

$$\frac{v_1 - v_0}{v_e} = -N \ln(s(1-l) + l).$$

Let the ratio of the payload mass, M_L, to the initial mass of the first stage, $M_0^{(1)}$ be λ:

$$\lambda = M_L/M_0^{(1)}.$$

(e) Show that the expression in (d) becomes

$$\frac{v_1 - v_0}{v_e} = -N \ln \left(s \left(1 - \lambda^{1/N} \right) + \lambda^{1/N} \right).$$

(f) Hence show that for large N the limit of improvement in the velocity increment due to multistaging is

$$\frac{v_1 - v_0}{v_e} = -(1-s) \ln(\lambda).$$

To show this you may find the following series useful:

$$a^x = e^{x \ln a} = 1 + x \ln a + \frac{(x \ln a)^2}{2!} + \frac{(x \ln a)^3}{3!} + \cdots \quad -\infty < x < \infty$$

$$\ln(1+x) = x - \frac{x^2}{2} + \frac{x^3}{3} - \frac{x^4}{4} + \cdots \quad -1 < x \le 1.$$

7.4 Workshop: Unit conversion

While schools ensure that students are only faced with one set of units (usually the SI units), laboratories are notoriously less fastidious in this matter. Accordingly it is vital that you are able to convert between different systems of units reliably. We shall illustrate two possible methods with two examples each.

(a) Convert 30 mph (miles per hour) into metres per second.
Solution 1: Convert the units themselves (1 miles = 1608 m; 1 h = 60 × 60 s):

$$30 \text{ mph} = \frac{30 \text{ miles}}{1 \text{ h}} = \frac{30 \times 1608 \text{ m}}{60 \times 60 \text{ s}} = \frac{30 \times 1608}{3600} \text{m/s} = 13.4 \text{ m/s}.$$

Solution 2: Keep multiplying by 1, since this cannot change the answer. Just find some very creative ways of writing the number 1, such as 1 = (1608 m/1 mile)

$$30 \text{ mph} = \frac{30 \text{ miles}}{1 \text{ h}} \times \frac{1 \text{ h}}{60 \times 60 \text{ s}} \times \frac{1608 \text{ m}}{1 \text{ miles}} = \frac{30 \times 1608}{3600} \text{m/s} = 13.4 \text{ m/s}.$$

Notice how the unwanted units cancel out if the working is laid out correctly.

(b) Convert 13.6 g/cm^3 into kilograms per cubic metre.
Solution 1: Convert the units themselves:

$$\frac{13.6 \text{ g}}{1 \text{ (cm)}^3} = \frac{13.6 \times 10^{-3} \text{kg}}{1 \times (10^{-2} \text{m})^3} = \frac{13.6 \times 10^{-3} \text{kg}}{1 \times 10^{-6} \text{m}^3} = 13.6 \times 10^3 \text{ kg/m}^3.$$

Solution 2: Keep multiplying by 1:

$$\frac{13.6 \text{ g}}{1 \text{ (cm)}^3} = \frac{13.6 \text{ g}}{1 \text{ (cm)}^3} \times \frac{10^{-3} \text{kg}}{1 \text{g}} \times \left(\frac{1 \text{ cm}}{10^{-2} \text{m}}\right)^3 = \frac{13.6 \times 10^{-3} \text{kg}}{1 \times 10^{-6} \text{m}^3}$$
$$= 13.6 \times 10^3 \text{ kg/m}^3.$$

Note that whenever prefixes are used in derived units, they are deemed to be stuck with glue to the unit on their immediate right, and should be squared or cubed together with it. Thus 1 cm^3 means 1 (cm)^3 = 10^{-6} m^3 *not* 1 c(m^3) = 10^{-2} m^3.

(c) Show that the acceleration due to gravity (9.81 m/s^2) is about 32 ft/s^2, where 1 ft is 0.305 m.

(d) Show that atmospheric pressure (1.03×10^5 N/m^2) is about 15 lbf/in.^{2}. One pound force (lbf) is the weight of one pound (1 lb = 0.454 kg) in the Earth's gravitational field (9.81 N/kg), while 1 in. is 25.4 mm.

(e) Some chemists still use the 'cgs' system of units, which is just like SI, except that mass is measured in grams and length is measured in centimetres. Show that the cgs unit of force (the dyne) is equal to 10^{-5} N, and that the cgs unit of energy (the erg) is equal to 10^{-7} J.

(f) Express the speed of light (3.00×10^8 m/s) in feet per nanosecond (ft/ns), where 1 ft = 0.305 m and 1 ns is 10^{-9} s.

144 *Miscellany*

7.5 Workshop: Dimensional analysis

When using SI units (kilograms, metres, seconds, amps, and so on), you will be used to using derived units wherever possible. If lengths are in metres, it seems perfectly natural to measure areas in square metres and volumes in cubic metres. If masses are in kilograms and time in seconds, it is no surprise that velocities are measured in m/s (or $\mathrm{m\,s^{-1}}$) and densities are measured in kg/m^3 (or $\mathrm{kg\,m^{-3}}$). Much about our quantities are given away by the units. The very fact that electric field strength is measured in N/C tells you that it must be related to some kind of force divided by some kind of charge, and thus gives you a very helpful hint as to what electric field strength is all about.

In addition, watching the units helps us get our equations right and guard against errors.

(a) By checking the units of left- and right-hand side of the following, determine which is the correct wave equation:

$$\frac{\partial^2 y}{\partial x^2} = \frac{1}{c^2}\frac{\partial^2 y}{\partial t^2} \quad \text{or} \quad \frac{\partial^2 y}{\partial t^2} = \frac{1}{c^2}\frac{\partial^2 y}{\partial x^2}.$$

Note: For the purposes of unit analysis ignore any differential 'd's (so dy/dt or $\partial y/\partial t$ reads as y/t and hence has units of m/s).*

Keeping tabs on the units is a very helpful way of not only checking our working, but also guessing likely forms of relationships.

(b) You are asked to find a relationship giving the time period of a simple pendulum. You think that it might depend on the mass m of the bob, the length L of the string, and the local gravitational field strength g. Assuming that the mass is in kilograms, the length in metres, and you express the field strength as an acceleration in m/s^2, how are you going to combine those quantities to produce a time in seconds?

Solution – empirical method: We wish to find the time in seconds. The only quantity to have seconds in it is g, so we start with this. The acceleration g is in m/s^2 (or $\mathrm{m\,s^{-2}}$), so to get seconds, we first have to divide it by some kind of length (e.g. L) and then we have a quantity measured in s^{-2}. To get seconds we put this to the power of $-1/2$. We have no need to use the mass in kilogram, and so the time period must be independent of the pendulum mass. This reasoning is summarized in the table below:

Quantity	Unit
g	$\mathrm{m\,s^{-2}}$
g/L	$\mathrm{m\,s^{-2}/m} = \mathrm{s^{-2}}$
$(g/L)^{-1/2} = \sqrt{L/g}$	$(\mathrm{s^{-2}})^{-1/2} = \mathrm{s}$

*And what about second derivatives? Again, we just ignore the 'd's. So $\partial^2 y/\partial t^2$ is read as y/t^2 and consequently has units of $\mathrm{m\,s^{-2}}$ – which is after all what you would expect for an acceleration. This is one way of looking at the reason why second derivatives are written d^2y/dt^2 and not dy^2/dt^2 – the quantity y on the top is most definitely not squared by the action of taking a second derivative.

Solution – analytical method: A more rigorous method is provided if we write the time period as a function $T \equiv m^\alpha L^\beta g^\gamma$, where α, β, and γ are unknown powers. We then write the equation in terms of its units:

$$\mathrm{s} \equiv \mathrm{kg}^\alpha \mathrm{m}^\beta \left(\frac{\mathrm{m}}{\mathrm{s}^2}\right)^\gamma.$$

The next stage is to equate the powers of the various units:

(i) Powers of kilograms $0 = \alpha$
(ii) Powers of metres $0 = \beta + \gamma$
(iii) Powers of seconds $1 = -2\gamma$,

and the simultaneous equations are easily solved to give $\alpha = 0$, $\beta = 1/2$, and $\gamma = -1/2$. So we know that T will be proportional to $m^0 L^{1/2} g^{-1/2} = \sqrt{L/g}$.

The full formula for the time period is $T = 2\pi \sqrt{L/g}$. Our methods got the last bit right, but notice that there was no way for us to guess the factor of 2π since it does not have units.

(c) Work out the form of the equation for the current in a wire I, expressed in terms of the cross-sectional area A, the charge on each electron q, the number of free electrons per cubic metre n, and the mean velocity of the electrons u.
(d) Show that if you know that centripetal acceleration is given by u^n/r, where u is the speed and r the radius of the turn, that the power n must be 2.
(e) Show that when farads (1 F = 1 C/V) are multiplied by ohms (1 Ω = 1 V/A), the result has the units of time. Is it then any wonder that if you double the resistance in an R–C circuit, you also double its time constant?

While these methods are useful, you may have noticed some complications. For example, if in question (b), we had given g in N/kg (as we normally do), things would have been more complicated. Hopefully you are already aware that 1 N, being the force that causes a 1 kg mass to have a 1 m/s² acceleration can also be written as 1 kg m/s², but it would none the less have made things more complicated.

We get round the problem by ensuring that we use only the SI base units (m, kg, s, A, mol, K, cd) when expressing the 'units' of quantities. This ensures that things work in a unique way since the base units expressed above are algebraically independent (none can be expressed in terms of the others).

However, some people prefer a slightly different way of working, called dimensional analysis, which builds upon the principles above. In this, we use square brackets [...] to denote 'the dimensions of', and use the symbols L, M, T, I, Θ to represent length, mass, time, electric current, and thermodynamic temperature, as in the table below. We can then write $[F] = [ma] = [MLT^{-2}]$, or $[E] = [mu^2] = [ML^2T^{-2}]$. Algebra can then be done amongst the dimensional quantities M, L, T, and so on. The square brackets remind us that we are only dealing with the dimensions, and that numerical factors (such as the $1/2$ in the expression for kinetic energy) are being ignored.

146 *Miscellany*

Quantity	SI Base unit	Dimension
Length	metre (m)	L
Time	second (s)	T
Mass	kilogram (kg)	M
Electric current	ampere (A)	I
Thermodynamic temperature	kelvin (K)	Θ
Amount of substance	mole (mol)	1
Luminous intensity	candela (cd)	no agreed symbol

The mole, being just a number of objects (usually molecules or atoms) has no dimensions – or more rigorously, it has the same dimensions as the number 1. Angles, when expressed in radians, are equally dimensionless – since the angle in radians is defined as the length of the arc of a circle (say, in metres) divided by the circle's radius in the same units. Thus angular frequency ω in rad/s has dimensions of T^{-1}, just like ordinary frequency (as measured in hertz).

(f) Express electric charge and voltage in terms of dimensions M, L, T, and I, and express their units (coulombs and volts) in terms of kg, m, s, and A. Thus show that the newton per coulomb (N/C) is identical numerically to the volt per metre (V/m), and so electric field strength measures the 'rate' of change of voltage with distance.

(g) Derive the likely form for the equation which gives the lift of an aircraft wing F in terms of the density of the air ρ, the cross-sectional area A of the wing (as presented to the airflow) and the speed of the air over the wing u.

The speed of fluid down a pipe u is related to the diameter of the pipe D, the length of the pipe L, the pressure difference across this length of pipe P, the density of the fluid ρ and the viscosity of the fluid μ (measured in N s/m² or kg m^{-1}s^{-1}). It is difficult to write an expression for P in terms of the other quantities, because there is more than one way of doing it. The approach is accordingly more complex than that outlined above.

(h) Using a method similar to the analytical solution to (b), suppose that the pressure difference is related to the other quantities by

$$[P] = [\rho^\alpha \mu^\beta D^\gamma L^\delta u^\varepsilon]$$

Given that we know the dimensions of all the quantities, show that we can set up equations for the unknown powers α–ε by equating the powers of the dimensions M, L, and T:

$$\begin{aligned} M \quad & 1 = \alpha + \beta \\ L \quad & -1 = -3\alpha - \beta + \gamma + \delta + \varepsilon \\ T \quad & 2 = \beta + \varepsilon. \end{aligned}$$

(i) The variables α and ε can be eliminated easily, and γ can also be expressed in terms of other variables without too much difficulty.* Once that is done show

---

*We could, of course, use the equations to eliminate β instead of α or δ instead of γ; however, the guidance here is given to enable the equation to end up in the form most familiar (and useful) for fluid mechanics problems.

that our equation reads

$$[P] = \left[\rho u^2 \left(\frac{\mu}{\rho u D}\right)^\beta \left(\frac{L}{D}\right)^\delta\right].$$

(j) It is found that the pressure difference is proportional to the length of the pipe. Show that once this is taken into account, we find $\delta = 1$, and

$$[P] = \left[\frac{\rho u^2 L}{D} \left(\frac{\mu}{\rho u D}\right)^\beta\right].$$

The fact that β can take any value (we could use *any* function of the term in round brackets) means that we cannot tie down the way in which the pressure drop will depend upon the viscosity using this method. However, it transpires that the reciprocal of the term in brackets $\rho u D/\mu$ is called the Reynold's number, and that its value determines the way in which the fluid flows down the pipe.

7.6 Workshop: Error analysis

No measurement is exact, and as such there is always a degree of uncertainty in a measured quantity. We can reduce the uncertainty by using a more precise measuring instrument or technique – but there always comes a point where there is no practical point in reducing the uncertainty further. While it might be interesting to make the measurement, there is no reason why the length of a curtain track needs to be known to the nearest 1 μm.

The business of science frequently requires us to know whether particular experiment agrees with theory – and for this we need to know the uncertainty of the experimental measurement. It is not good rejecting a wonderful theory on the grounds that its prediction for the mass of the electron is out by 1.2×10^{-32} kg if the electronic mass is only known to the nearest 7×10^{-32} kg.

In engineering, the use of uncertainties is even more important. When designing a supporting column for a multi-storey car park we need to take into account that different castings of the same concrete recipe can (and do) have different strengths, and that the projected weight of the building above might be exceeded if the car park ends up full of 4 × 4 s. We can only make a safe choice of column thickness if we are aware of these issues.

We start by using 'absolute uncertainties'. If a current is 1.3 ± 0.1 A, then the absolute uncertainty is 0.1 A. Putting it crudely, you can expect the measurement to usually be within 0.1 A of 1.3 A.

Absolute uncertainties are usually estimated in one of two ways – from the measuring instrument, or from the data when more than one reading has been taken. If your metre ruler is calibrated in millimetres, and you use it carefully, it will probably make sense to say that the absolute uncertainty is 0.5 mm (half of the smallest division). Equally well, you may be told that a particular ammeter is only accurate to 0.02 A.

On the other hand, you might make the measurement a number of times, and infer the uncertainty from the spread of results. If you measured the force required to stretch a spring by 15 cm three times, and got the answers 6.4, 6.5, and 6.3 N, you would usually be justified in estimating the uncertainty as 0.1 N, since the readings are all within 0.1 N of the central value. With a large amount of data, you might wish to use the standard deviation to calculate a more formal measure of the uncertainty – but this is not always required.

The *relative uncertainty* in a measurement is the absolute uncertainty expressed as a percentage of the measurement itself.

$$\text{Relative uncertainty (\%)} = \frac{\text{Absolute uncertainty}}{\text{Measurement}} \times 100\%,$$

where the absolute uncertainty and the measurement must be expressed in the same units.

(a) Calculate the relative uncertainty of the following measurements:
   (i) 3.03 ± 0.2 m
   (ii) 2.34 m ± 2 cm
   (iii) 24.3 km ± 240 m
   (iv) 1.602×10^{-19} C $\pm 3 \times 10^{-22}$ C.

(b) Work out the relative and absolute uncertainty (to one significant figure) of the following set of voltage measurements: 34.2, 36.2, 35.2, 35.8, and 36.1 mV.

(c) A house is 7.5 ± 0.2 m high, and a TV aerial is 2.1 ± 0.1 m tall. If the aerial is put onto the top of the house, how high is the top of the aerial above the ground?

(d) My mass, as measured by bathroom scales, is 63.2 ± 0.2 kg. When holding my daughter, it is 74.7 ± 0.2 kg. What is the mass of my daughter?

Hopefully you got the answers 9.6 ± 0.3 m and 11.5 ± 0.4 kg, respectively. In (c) the worst case scenario is if both measurements are underestimates (or both are overestimates) – in which case the errors add to 0.3 m. In (d) the situation is reversed. The worst errors occur if one measurement is too small while the other is too big. Accordingly the biggest mass allowable by the data is $74.9 - 63.0 = 11.9$ kg, which is 0.4 kg different from the expected result. In either case, you would be wrong to make the assumption that the errors would cancel out. Of course they might, but then again they might not, and you would better not take the gamble.

These exercises give us the following rule: *When adding or subtracting measurements, you ADD the absolute uncertainties.*

When dealing with a large amount of data, it is highly unlikely that *all* of the random errors will be overestimates. Accordingly, the 'statistically correct' thing to do is not simply to add the absolute uncertainties, but to square them, then add them, then take the square root. This allows for the possibility of some of the errors cancelling out. This is sometimes called combining errors in quadrature.

(e) Using the method of adding errors in quadrature, show that expected absolute uncertainties in (c) and (d) are ±0.22 m and ±0.28 kg (to 2 significant figures).

(f) A 10 cm long rod is attached to a 20 cm rod with a hinge. The angle between the two rods at the hinge is equally likely to take any value. Make an estimate of the most likely distance between the two free ends of the rods.

(g) Nine measurements are made of the magnetic field strength of the Earth. Each measurement carries an uncertainty of 1.0×10^{-6} T. Using the method of adding errors in quadrature, what absolute uncertainty would you expect for (1) the sum and (2) the mean of the nine measurements.

Hopefully you found in (g) that your absolute error for the sum was $\sqrt{9 \times (1.0 \times 10^{-6})^2} = 3.0 \times 10^{-6}$ T, and thus the absolute error for the mean (on division by 9) was 3.3×10^{-7} T. This is the justification for expecting a mean of many measurements to be less uncertain than a single measurement. You expect some of the overestimates to cancel out underestimates when they are added. Crudely put, we expect:

$$\text{Uncertainty of mean} = \frac{\text{Uncertainty of each measurement}}{\sqrt{\text{Number of measurements}}}.$$

(h) A journey is 87.3 ± 0.5 km long, and I travel along the road at a speed of 92 ± 1 km/h. Calculate (1) the expected time for the journey, (2) the longest time the journey could take and then (3) compare the relative uncertainty of the time with the relative uncertainties of the distance and speed.

Part (h) brings us onto our next approximate rule: *When measurements are multiplied or divided, the relative error of the result equals the SUM of the relative uncertainties of the individual measurements.*

(i) If the voltage across a thermistor is 6.43 V ± 30 mV, and the current flowing through it is 7.5 ± 0.1 mA, what is the resistance of the thermistor? Give your answer with a relative uncertainty.

(j) Calculate the relative uncertainty in the kinetic energy of a 1.000 kg mass travelling at a speed of 3.4 ± 0.3 m/s. How does this compare with the relative uncertainty of the speed?

Our third rule is that if the relative uncertainty of x is $p\%$, the relative uncertainty of x^n is $np\%$. Notice that the relative uncertainty in x^{-1} is the same as that in x, but of opposite sign (underestimates become overestimates and vice versa).

Our second and third rules can be justified and extended to other situations using the chain rule in calculus. If $f(x)$ is some function of a measurement x (which has absolute uncertainty δx), then the absolute error in $f(x)$ is

$$f(x + \delta x) - f(x) \approx \frac{df}{dx} \delta x,$$

while the relative uncertainty is

$$\frac{f(x + \delta x) - f(x)}{f(x)} \times 100\% \approx \frac{1}{f} \frac{df}{dx} \delta x \times 100\%.$$

(k) Use the equation for relative uncertainty to show that if $f(x) = x^n$, then the relative uncertainty in $f(x)$ is n times the relative uncertainty in x.

If a function depends on more than one measurement, then the total uncertainty is the sum of the uncertainties due to the dependence on each of the measurements:

$$f(x + \delta x,\ y + \delta y) - f(x, y) \approx \frac{\partial f}{\partial x} \delta x + \frac{\partial f}{\partial y} \delta y,$$

$$\frac{f(x + \delta x,\ y + \delta y) - f(x, y)}{f(x, y)} \approx \frac{1}{f}\frac{\partial f}{\partial x} \delta x + \frac{1}{f}\frac{\partial f}{\partial y} \delta y.$$

Note the use of partial derivatives to show that in each case we are differentiating with respect to one of the variables holding the other constant.

(l) Use the expression above to show that the relative error in the function $f(x, y) = xy$ is equal to the relative error in x added to the relative error in y.

7.7 Workshop: Centres of mass

In the two-body problem, introduced in Section 3.2.1, the location of the centre of mass may be obtained by finding the vector $\mathbf{R}$ that satisfies the following expression (Figure 7.3):

$$m_1 \mathbf{r}_1 + m_2 \mathbf{r}_2 = (m_1 + m_2)\mathbf{R}.$$

Here O is some arbitrary origin. Generalizing this for any number of point masses m_i of position vectors $\mathbf{r}_i$ with respect to some arbitrary origin we have

$$\sum_i m_i \mathbf{r}_i = \left(\sum_i m_i\right) \mathbf{R}.$$

This expression offers us a clue as to how we might locate the centre of mass of a rigid body. In the case of a rigid body, the number of 'particles' is effectively infinite, though of course we are not saying here that we would perform a sum over the atoms

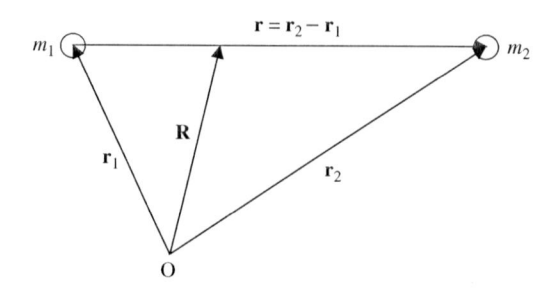

Fig. 7.3

Workshop: Centres of mass 7.7

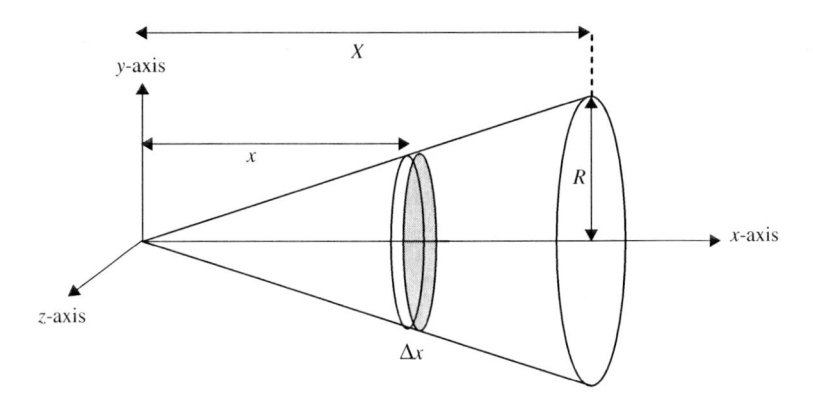

Fig. 7.4

of the body. The usual approach is to imagine the rigid body to be made up of lots of infinitesimally small pieces, the summation over these pieces converges to a finite sum, which turns out to be an integral. For this reason, it might be informative to have a go at workshop 7.1 before proceeding to the problem in this section.

Let us apply this procedure to locating the centre of mass of a uniform solid right circular cone of total mass M. With the cone material being uniform, we have a constant density throughout, which we shall call here ρ. However, we can of course extend this to include densities that also vary with position.

The cone in Figure 7.4 is cylindrically symmetric about the x-axis. Therefore the centre of mass must lie along this axis. We now imagine that the cone is in fact made up of lots of infinitesimally thin discs each of thickness Δx. The centre of mass of each of these discs is located at its centre, so the location of the centre of mass of the cone is a vector $\mathbf{R}_{\text{cm}}$ that satisfies the following expression:

$$\sum_i \Delta m_i \mathbf{r}_i = \left(\sum_i \Delta m_i \right) \mathbf{R}_{\text{cm}},$$

where Δm_i is the mass of ith disc and $\mathbf{r}_i$ is the location of the centre of the ith disc.

(a) Show that

$$\mathbf{R}_{\text{cm}} = \left(\frac{\rho \pi R^2}{M X^2} \int_0^X x^3 dx \right) \hat{\mathbf{x}},$$

and perform the integral to verify that the centre of mass of a uniform solid right circular cone is located $1/4$ of the cone's height above the base of the cone.

(b) Now assume that the density of the cone in Figure 7.4 is a function of y:

(i) For $0 \leq y \leq R$, $\rho = 2\rho_0$
(ii) For $-R \leq y < 0$, $\rho = \rho_0$

and show that the centre of mass must now lie on the line $y = \dfrac{4R}{9X\pi}x$.

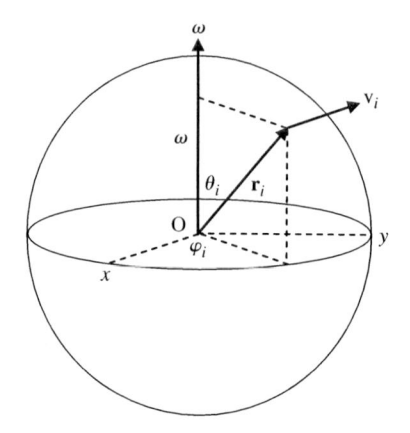

Fig. 7.5

7.8 Workshop: Rigid body dynamics

In the section on orbits (3.2), it became apparent that the form of the gravitational interaction and associated equations of motion led to some important quantities. One of these was the vector $\mathbf{h}$:

$$\mathbf{h} = \mathbf{r} \times \mathbf{v}.$$

This vector was a constant of the motion if the vector $\mathbf{r} \times \mathbf{a} = \mathbf{0}$, as

$$\mathbf{r} \times \mathbf{a} = \mathbf{0} \Rightarrow \frac{d}{dt}(\mathbf{r} \times \mathbf{v}) = \mathbf{0},$$

which of course is true for the gravitational interaction as $\mathbf{a}$ and $\mathbf{r}$ are parallel.

In Figure 7.5 we have a rigid body (a sphere in this case, but all this applies to any rigid body), centred on the origin of the coordinate system, rotating about the z-axis. A rigid body may be thought of a collection of lots of little masses that are rigidly held together. For each of these masses the angular velocity $\boldsymbol{\omega}$ of the rotation about the z-axis is the same, but the instantaneous velocity will be different:

$$\mathbf{v}_i = \boldsymbol{\omega} \times \mathbf{r}_i,$$

where i denotes the ith particle. We can of course define an $\mathbf{h}_i$ for each of these particles. If we multiply $\mathbf{h}_i$ by the mass of each of the particles and sum over all the particles we construct a vector, say $\mathbf{L}$, which is parallel to the z-axis.

$$\mathbf{L} = \sum_i \mathbf{r}_i \times m_i \mathbf{v}_i.$$

Now $m_i \mathbf{v}_i$ is in fact the instantaneous linear momentum of the ith particle and each of the $\mathbf{r}_i \times m_i \mathbf{v}_i$ is usually referred to as the *angular momentum* of the ith particle, so the vector **L** is the total angular momentum of the rigid body. Physicists find this vector useful because its rate of change is something that is quite familiar:

$$\frac{d}{dt}\mathbf{L} = \sum_i \frac{d}{dt}(\mathbf{r}_i) \times m_i\mathbf{v}_i + \mathbf{r}_i \times \frac{d}{dt}(m_i\mathbf{v}_i) = \sum_i \mathbf{r}_i \times \mathbf{F}_i,$$

where $\mathbf{F}_i$ is the force on the ith particle. The object on the far right is called the *torque* on the rigid body (with each of the $\mathbf{r}_i \times \mathbf{F}_i$ being the torque on each of the ith particles). It is a twisting force, which is of course why $\mathbf{r} \times \mathbf{v}$ is a constant for the Kepler problem as the gravitational interaction cannot apply a twist (the force is always through the centres of mass of the interacting bodies). Therefore we have a parallelism between rotational dynamics and linear dynamics.

In linear dynamics the rate of change of momentum is equal to force:

$$\mathbf{F} = \frac{d}{dt}\mathbf{p},$$

so when the resultant force on the system is zero then $(d/dt)\mathbf{p} = \mathbf{0}$, and $\mathbf{p}$, the total momentum (the vector sum of component momenta) of the system must be a constant.

In rotational dynamics the rate of change of angular momentum is equal to torque – a force is applied at a point on the body and the torque is just the force multiplied by the perpendicular distance from the axis of rotation to the point of application. This is just $\mathbf{r} \times \mathbf{F}$ ($Fr\sin\theta\hat{\mathbf{z}}$ has indeed a magnitude that is just the perpendicular distance of the axis from the point of application multiplied by the force).

One more thing,*

$$\mathbf{r} \times \mathbf{F} = \frac{d}{dt}(\mathbf{r} \times \mathbf{p}),$$

so when the resultant torque on the system is zero then $(d/dt)(\mathbf{r} \times \mathbf{p}) = \mathbf{0}$, which means that $(\mathbf{r} \times \mathbf{p})$, the angular momentum of the system must be a constant. This is merely a statement of the conservation of angular momentum in systems where there is no resultant torque.

(a) For a rigid body rotating about some arbitrary axis, the angular velocity will take the form:

$$\boldsymbol{\omega} = \begin{pmatrix} \omega_x \\ \omega_y \\ \omega_z \end{pmatrix}.$$

---

*You can easily see this by differentiating $(\mathbf{r} \times \mathbf{p})$ and respecting the order of the vector product; that is, $d/dt(\mathbf{r} \times \mathbf{p}) = (d/dt)\mathbf{r} \times \mathbf{p} + \mathbf{r} \times (d/dt)\mathbf{p} = \mathbf{v} \times \mathbf{p} + \mathbf{r} \times \mathbf{F} = \mathbf{r} \times \mathbf{F}$ as **v** is parallel to **p** and the vector product of parallel vectors is zero.

154 Miscellany

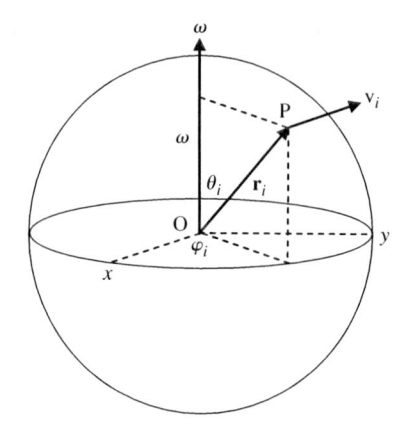

Fig. 7.6

Show that $\mathbf{L} = \sum_i \mathbf{r}_i \times m_i \mathbf{v}_i = \sum_i \mathbf{r}_i \times m_i (\boldsymbol{\omega} \times \mathbf{r}_i) =$

$$\begin{pmatrix} \sum_i m_i(r_i^2 - x_i^2) & \sum_i -m_i x_i y_i & \sum_i -m_i x_i z_i \\ \sum_i -m_i x_i y_i & \sum_i m_i(r_i^2 - y_i^2) & \sum_i -m_i y_i z_i \\ \sum_i -m_i x_i z_i & \sum_i -m_i y_i z_i & \sum_i m_i(r_i^2 - z_i^2) \end{pmatrix} \boldsymbol{\omega}.$$

Hint: Section 3.1.5.

The matrix here is called the *moment of inertia* and is usually given the symbol $\underline{I}$ (here the underline denotes that it is a 3×3 matrix).

(b) The sphere in our example (Figure 7.6) may be broken up into little pieces each of volume:

$$\Delta V = r^2 \sin\theta \, \Delta r \Delta\theta \Delta\varphi,$$

where r $(0 \leq r \leq R)$, θ $(0 \leq \theta \leq \pi)$, and φ $(0 \leq \varphi \leq 2\pi)$ refer to the coordinates of the little piece (say $\mathbf{P}$) in *spherical polar coordinates*. The second half of workshop in 7.1 introduces this coordinate system and it might be a good idea to have a look at that before attempting the rest of this workshop.

Assuming the sphere is uniform and is rotating about the z-axis only, which passes through its centre, show that the matrix expression in (a) collapses to

$$\mathbf{L} = I_{zz}\omega_z \hat{\mathbf{z}},$$

where $I_{zz} = \rho \iiint r^4 \sin^3\theta \, dr \, d\theta \, d\varphi$, with ρ being the density of the material.

(c) Perform the integral and show that

$$I_{zz} = \rho \iiint r^4 \sin^3\theta \, dr \, d\theta \, d\varphi = \frac{8}{15}\pi\rho R^5,$$

where R is the radius of the sphere. You may find the following indefinite integral helpful:

$$\int \sin^3 \theta d\theta = -\cos\theta + \frac{\cos^3\theta}{3}.$$

For symmetric bodies that only rotate about one of the coordinate axes, the moment of inertia collapses to a single number.

(d) The total kinetic energy of a rotating rigid body is of course the sum of all the kinetic energies of all the little masses that make up the body:

$$K = \sum_i \frac{1}{2} m_i \mathbf{v}_i \cdot \mathbf{v}_i,$$

Show that $K = \frac{1}{2}\boldsymbol{\omega} \cdot \boldsymbol{I}\boldsymbol{\omega}$, which collapses to

$$K = \frac{1}{2} I_{zz} \omega^2 = \frac{1}{2} \frac{8}{15} \pi \rho R^5 \omega^2$$

for our sphere.

The object I, the moment of inertia, describes the mass distribution in the rigid body.

(e) Our sphere rolls down an inclined plane, rough enough to prevent slipping. The frictional force F at the point of contact applies a torque to the sphere and as it accelerates down the slope (maintaining contact with the slope) the rotation picks up speed as well. Show that F is given by:

$$F = \frac{2}{5} MR \frac{d\omega}{dt},$$

where M is the total mass of the sphere.

With no slipping the speed V of the sphere along the plane is just:

$$V = R\omega.$$

Show that the acceleration of the centre of mass of the sphere is given by

$$a = \frac{5}{7} g \sin\alpha,$$

where g is the acceleration due to gravity and α is the inclination of the plane.

7.9 Workshop: Parallel axes theorem

This workshop follows on from the previous one (Section 7.8). Indeed, the parallel axes theorem refers to the transformation of the moment of inertia if the axis of rotation is shifted to another axis that is parallel to the first. You are advised to have a go at the previous workshop (or at least look over the solutions) before attempting this one.

156 Miscellany

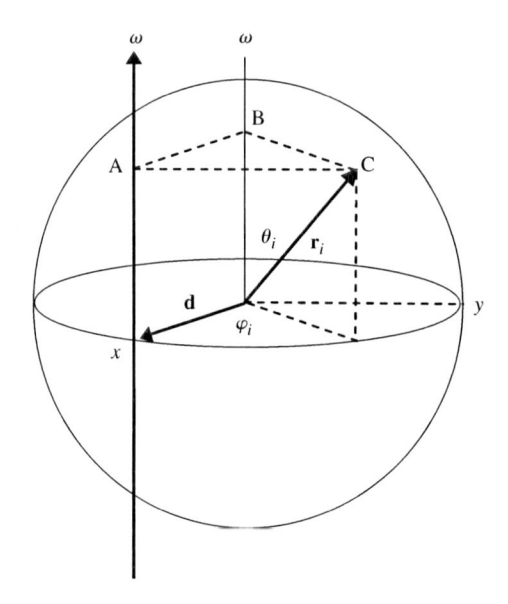

Fig. 7.7

In the previous workshop, the angular momentum **L** of the sphere about the z-axis was given by

$$\mathbf{L} = I_{zz}\omega_z\hat{\mathbf{z}},$$

where $I_{zz} = \rho \iiint r^4 \sin^3\theta \, dr d\theta d\varphi$, with ρ being the density of the material, and ω_z is the magnitude of the angular velocity about the z-axis. The expression for I_{zz} looks complicated, but is really only a term of the form:

$$I_{zz} = \iiint dM \, r^2 \sin^2\theta,$$

or, in other words, sum up all the little masses dM multiplied by the square of their distances away from the z-axis ($r\sin\theta$).

(a) If we shift the axis of rotation to the new axis (see Figure 7.7) show that I_{zz} transforms to

$$I_{zz} = \rho \iiint (r^2 \sin^2\theta + d^2) r^2 \sin\theta \, dr \, d\theta \, d\varphi,$$

where d is the magnitude of the displacement **d** (look at the triangle ABC). This expression becomes:

$$I'_{zz} = I_{zz} + Md^2.$$

With d coming out of the integral signs because it is a constant, we can see that this is a general result; that is, the moment of inertia about a new axis that is parallel to the old is just the old moment of inertia plus Md^2 (the mass of the rigid body multiplied by the square of the distance between the new axis and the old).

7.10 Workshop: Perpendicular axes theorem

Like the previous workshop, this section follows on from Section 7.8. You are advised to have a go at Section 7.8 (or at least look over the solutions) before attempting this one.

The sphere in Figure 7.8 is rotating about an axis through its centre. The angular velocity of this rotation is

$$\boldsymbol{\omega} = \omega_x \hat{\mathbf{x}} + \omega_y \hat{\mathbf{y}} + \omega_z \hat{\mathbf{z}} = \begin{pmatrix} \omega_x \\ \omega_y \\ \omega_z \end{pmatrix}.$$

In Section 7.8, we discovered that the angular momentum $\mathbf{L}$ of a rigid body rotating at an angular velocity ω is given by

$$\mathbf{L} = \begin{pmatrix} \sum_i m_i(r_i^2 - x_i^2) & \sum_i -m_i x_i y_i & \sum_i -m_i x_i z_i \\ \sum_i -m_i x_i y_i & \sum_i m_i(r_i^2 - y_i^2) & \sum_i -m_i y_i z_i \\ \sum_i -m_i x_i z_i & \sum_i -m_i y_i z_i & \sum_i m_i(r_i^2 - z_i^2) \end{pmatrix} \boldsymbol{\omega},$$

where the rigid body has been imagined to be made of a collection of lots of little masses that are rigidly held together. The letter i denotes the ith little mass in the collection and the summation is over the whole collection of masses.

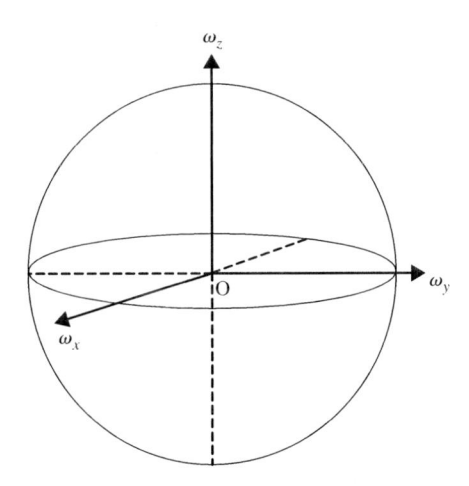

Fig. 7.8

158 Miscellany

The sphere in this example is uniform so terms like

$$\sum_i -m_i x_i y_i$$

will be zero because for a given value of x_{io} there will be two terms:

$$m_i x_{io} y_i + m_i x_{io}(-y_i) = 0,$$

and similarly for a y_{io}:

$$m_i x_i y_{io} + m_i (-x_{io}) y_{io} = 0.$$

So, **L** becomes

$$\mathbf{L} = \begin{pmatrix} I_{xx} & 0 & 0 \\ 0 & I_{yy} & 0 \\ 0 & 0 & I_{zz} \end{pmatrix} \boldsymbol{\omega}.$$

The term $I_{zz} = \sum_i m_i (r_i^2 - z_i^2) = \sum_i m_i (x_i^2 + y_i^2)$ just means the following: Sum up all the little masses multiplied by the square of the distance from the z-axis. The terms $\sum_i m_i (r_i^2 - x_i^2)$ and $\sum_i m_i (r_i^2 - y_i^2)$ are just the same only the relevant axes are, respectively, the x and y axes. The symmetry and uniformity of our sphere mean that

$$I_{xx} = I_{yy} = I_{zz} = \frac{2}{5} MR^2,$$

where M and R are, respectively, the mass and radius of the sphere.

Let us apply all this to the uniform disc in Figure 7.9. Let the mass per unit area be σ.

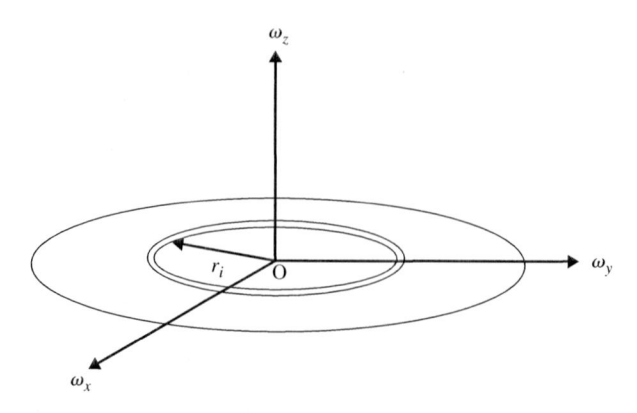

Fig. 7.9

(a) Show that for each ring of radius r_i and thickness Δr, $I_{zz} = (2\pi r_i \Delta r \sigma) r_i^2$.
(b) I_{zz} for the disc is the sum of all the little ring masses. Show that this comes out to be

$$I_{zz} = \frac{1}{2} MR^2,$$

where M and R are, respectively, the mass and radius of the disc.

(c) Now each $r_i^2 = x_i^2 + y_i^2$, so $\sum_i m_i r_i^2 = \sum_i m_i (x_i^2 + y_i^2)$. Hence show that if $I_{zz} = \frac{1}{2} MR^2$, then $I_{xx} = I_{yy} = \frac{1}{4} MR^2$, and that the angular momentum of our disc is

$$\mathbf{L} = \begin{pmatrix} \frac{1}{4}MR^2 & 0 & 0 \\ 0 & \frac{1}{4}MR^2 & 0 \\ 0 & 0 & \frac{1}{2}MR^2 \end{pmatrix} \begin{pmatrix} \omega_x \\ \omega_y \\ \omega_z \end{pmatrix},$$

In general, if the moments of inertia of a lamina about two perpendicular axes in its plane which meet at O are I_{xx} and I_{yy}, then the moment of inertia about an axis through O perpendicular to the plane of the lamina is

$$I_{zz} = I_{xx} + I_{yy}.$$

7.11 Workshop: Orbital energy and orbit classification

It is informative to have an idea of the physical meaning of the constant scalar in (3.61) of Section 3.2.2. To do this we first consider our system of two gravitationally interacting masses m_1 and m_2 from the point of view of the centre of mass. In Figure 7.10, O is an arbitrary origin of coordinates and C is the centre of mass of the system. Notice that we have attached the rotating coordinates $(\hat{\mathbf{x}}', \hat{\mathbf{y}}')$ to C.

The total energy E of this system, referred to the centre of mass coordinates, is given by

$$E = \frac{1}{2} m_1 V_1^2 + \frac{1}{2} m_2 V_2^2 - \frac{Gm_1 m_2}{r},$$

where

$$\mathbf{V}_1 = -\frac{dR_1}{dt} \hat{\mathbf{x}}' - \omega R_1 \hat{\mathbf{y}}' \quad \text{and} \quad \mathbf{V}_2 = \frac{dR_2}{dt} \hat{\mathbf{x}}' + \omega R_2 \hat{\mathbf{y}}',$$

$\mathbf{V}_1 \cdot \mathbf{V}_1 = V_1^2$, $\mathbf{V}_2 \cdot \mathbf{V}_2 = V_2^2$ (see Figure 7.10), and $\mathbf{r} = \mathbf{r}_2 - \mathbf{r}_1 = r\hat{\mathbf{x}}'$.

(a) Show that $R_1 = \dfrac{m_2}{(m_1 + m_2)} r$ and $R_2 = \dfrac{m_1}{(m_1 + m_2)} r$.

(b) Now show that $\dfrac{1}{2} m_1 V_1^2 + \dfrac{1}{2} m_2 V_2^2 = \dfrac{1}{2} \dfrac{m_1 m_2}{(m_1 + m_2)} v^2$, where $v^2 = \dfrac{d}{dt}\mathbf{r} \cdot \dfrac{d}{dt}\mathbf{r}$.

160   Miscellany

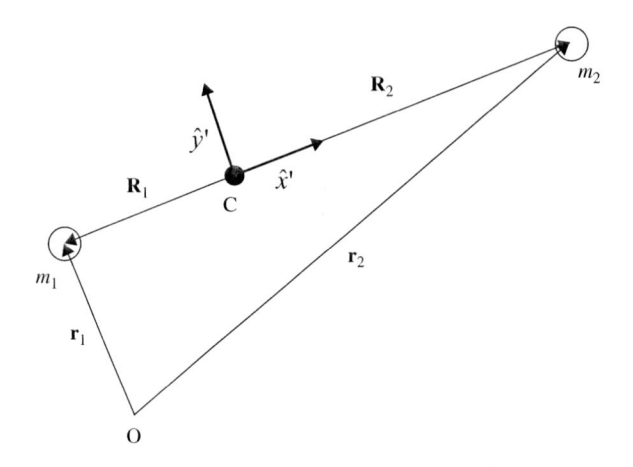

Fig. 7.10

(c) Hence show that

$$\frac{v^2}{2} - \frac{G(m_1+m_2)}{r} = \frac{E}{\mu},$$

where $\mu = (m_1 m_2)/(m_1 + m_2)$, which is often referred to as the *reduced mass*.*
This of course means that the constant scalar in (3.61) is effectively a statement of the *conservation of energy*.

(d) Determine $\mathbf{v} \times \mathbf{h}$ with $\mathbf{v} = (dr/dt)\hat{\mathbf{x}}' + r\omega\hat{\mathbf{y}}'$ and $\mathbf{h} = \omega^2 r \hat{\mathbf{z}}'$.

(e) Show that $(\mathbf{v} \times \mathbf{h}) \cdot (\mathbf{v} \times \mathbf{h}) = v^2 h^2$, where $h = \omega^2 r$.
From (3.51), $(\mathbf{v} \times \mathbf{h}) = G(m_1 + m_2)(\hat{\mathbf{x}}' + \mathbf{e})$, so

$$(\mathbf{v} \times \mathbf{h}) \cdot (\mathbf{v} \times \mathbf{h}) = G^2(m_1+m_2)^2(1 + 2e\cos\theta + e^2),$$

where θ is the angle between the unit vector $\hat{\mathbf{x}}'$ and the constant vector $\mathbf{e}$. Hence

$$v^2 = \frac{G^2(m_1+m_2)^2}{h^2}(1 + 2e\cos\theta + e^2).$$

---
*Reduced mass is the 'effective' mass appearing in the two-body problem. This is a quantity with the units of mass, which allows the two-body problem to be solved as if it were a one-body problem. For the two-body problem, $\mathbf{F}_1 + \mathbf{F}_2 = m_1\mathbf{a}_1 + m_2\mathbf{a}_2 = \mathbf{0}$ is a statement of the system's isolation from all other bodies (and hence a statement of Newton's third law). This expression can of course be rearranged to:

$$\mathbf{a} = \mathbf{a}_1 - \mathbf{a}_2 = (1 + \tfrac{m_2}{m_1})\mathbf{a}_1 = \frac{\mathbf{F}_1}{\mu}; \quad \text{that is,}$$

mass 1 moves with respect to mass 2 as a body equal in mass to the reduced mass μ.

(f) With
$$r = \frac{h^2/(G(m_1+m_2))}{(1+e\cos\theta)},$$
show that $\dfrac{v^2}{2} - \dfrac{G(m_1+m_2)}{r} = \dfrac{G^2(m_1+m_2)^2}{2h^2}(e^2-1).$

The eccentricity (e) of the trajectory is related directly to the total energy (E) of the system. The orbit is effectively classified by the sign of the total energy.

For:
(i) $0 \leq e < 1$ the total energy is negative ($E < 0$) so it must be a bound orbit and hence we have either an elliptical or circular orbit (*NB*: The circular orbit is the lowest energy orbit).
(ii) $e = 1$ the total energy is zero ($E = 0$), and we have just the conditions for escape. This is a parabolic orbit.
(iii) $e > 1$ the total energy is positive ($E > 0$). We have a hyperbolic orbit the bodies escape from each other's gravitation and still have positive kinetic energy at an infinite distance from each other.

8
Summary of equations

8.1 Linear mechanics

Fundamental theorems of calculus

$$\frac{df}{dt} = \lim_{\Delta t \to 0} \left\{ \frac{f(t + \Delta t) - f(t)}{\Delta t} \right\}$$

$$f(t) = \frac{d}{dt} \int_{t_0}^{t} f(t') dt' = \int_{t_0}^{t} \frac{df}{dt'} dt' + f(t_0)$$

Applied to the scalar components of acceleration (a), velocity (v), and displacement (s)

$$v(t) = \int_{t_0}^{t} \frac{dv}{dt'} dt' + v(t_0) = \int_{t_0}^{t} a(t') dt' + v(t_0)$$

$$s(t) = \int_{t_0}^{t} \frac{ds}{dt'} dt' + s(t_0) = \int_{t_0}^{t} v(t') dt' + s(t_0)$$

$$s(t) = \int_{t_0}^{t} \left(\int_{t_0}^{t'} a(t'') dt'' + v(t_0) \right) dt' + s(t_0)$$

$$= \int_{t_0}^{t} \left(\int_{t_0}^{t'} a(t'') dt'' \right) dt' + v(t_0)(t - t_0) + s(t_0)$$

For a uniform acceleration a

$$s(t) = s(t_0) + v(t_0)(t - t_0) + \frac{1}{2} a(t - t_0)^2$$

Vector equations of motion for velocity $\mathbf{v}$, and displacement $\mathbf{r}$ under uniform acceleration $\mathbf{a}$

$$\mathbf{v}(t) = \mathbf{v}(t_0) + \mathbf{a}(t - t_0)$$

$$\mathbf{r}(t) - \mathbf{r}(t_0) = \frac{(\mathbf{v}(t_0) + \mathbf{v}(t))(t - t_0)}{2} = \mathbf{v}(t_0)(t - t_0) + \frac{1}{2} \mathbf{a}(t - t_0)^2$$

Resultant vector **R** and relative vectors $\mathbf{r}_{21}$ and $\mathbf{r}_{12}$

$$\mathbf{R} = \mathbf{r}_1 + \mathbf{r}_2$$
$$\mathbf{r}_{21} = \mathbf{r}_2 - \mathbf{r}_1 = -\mathbf{r}_{12}$$

Momentum and resultant force **F**

$$\mathbf{p} = m\mathbf{v}$$
$$\mathbf{F} = \frac{d}{dt}\mathbf{p} = \frac{d}{dt}(m\mathbf{v}) = m\frac{d\mathbf{v}}{dt} + \mathbf{v}\frac{dm}{dt}$$

Conservation of momentum in a closed system

$$\frac{d}{dt}\sum_i \mathbf{p}_i = \mathbf{0} \Rightarrow \sum_i \mathbf{p}_i = \mathbf{P}_{\text{constant}}$$

Impulse $\Delta\mathbf{p}$

$$\int_{t_0}^{t} \mathbf{F}(t')dt' = \int_{t_0}^{t} \frac{d\mathbf{p}}{dt'}dt' = \mathbf{p}(t) - \mathbf{p}(t') = \Delta\mathbf{p}$$

Newton's experimental law

$$v_2 - v_1 = -e(u_2 - u_1)$$

8.2 Fields

Scalar product of vectors

$$\mathbf{a}\cdot\mathbf{b} = \mathbf{b}\cdot\mathbf{a} = ab\cos\theta$$
$$\begin{pmatrix} a_x \\ a_y \\ a_z \end{pmatrix} \cdot \begin{pmatrix} b_x \\ b_y \\ b_z \end{pmatrix} = a_x b_x + a_y b_y + a_z b_z$$

Equation linking acceleration, velocity, and displacement

$$v^2 = u^2 + 2\mathbf{a}\cdot\mathbf{s}$$

Motion in gravitational fields

Field strength (N/kg) = Gravitational force (N) ÷ Mass (kg) $\mathbf{g} = \mathbf{F}/m$

	1-D	3-D	Non-uniform fields
Work done by field	mgy	$m\mathbf{g}\cdot\mathbf{s}$	$\int m\mathbf{g}\cdot d\mathbf{s}$
Kinetic energy	$\frac{1}{2}mv^2 = \frac{1}{2}m\mathbf{v}\cdot\mathbf{v}$		
Change in potential energy	mgh	$-m\mathbf{g}\cdot\mathbf{s}$	$-\int m\mathbf{g}\cdot d\mathbf{s}$
Change in potential = potential energy per kg	gh	$-\mathbf{g}\cdot\mathbf{s}$	$-\int \mathbf{g}\cdot d\mathbf{s}$

Motion in electrostatic fields

Field strength (N/C) = Electrostatic force (N) ÷ Charge (C) $\mathbf{E} = \mathbf{F}/q$

	1-D	3-D	Non-uniform fields
Work done by field	qEy	$q\mathbf{E}\cdot\mathbf{s}$	$\int q\mathbf{E}\cdot d\mathbf{s}$
Kinetic energy	$\frac{1}{2}mv^2 = \frac{1}{2}m\mathbf{v}\cdot\mathbf{v}$		
Change in potential energy	qEh	$-q\mathbf{E}\cdot\mathbf{s}$	$-\int q\mathbf{E}\cdot d\mathbf{s}$
Change in potential = potential energy per C	Eh	$-\mathbf{E}\cdot\mathbf{s}$	$-\int \mathbf{E}\cdot d\mathbf{s}$

Gradient function

$$\nabla\phi = \begin{pmatrix} \frac{\partial\phi}{\partial x} \\ \frac{\partial\phi}{\partial y} \\ \frac{\partial\phi}{\partial y} \end{pmatrix} \qquad \mathbf{g}(or\,\mathbf{E}) = -\nabla\phi$$

Setting up fields

'Number of field lines' leaving a region defined as, for example, $\oiint_S \mathbf{E}\cdot d\mathbf{S}$, where $\mathbf{S}$ is a surface that bounds the region.

	Equation representing number of field lines
Electrostatic	$\iint_S \mathbf{E} \cdot d\mathbf{S} = \dfrac{Q}{\varepsilon_0}$
Gravitational	$\iint_S \mathbf{g} \cdot d\mathbf{S} = -4\pi MG$
Magnetic	$\iint_S \mathbf{B} \cdot d\mathbf{S} = 0$

Particular expressions for field strength

Electrostatic field due to point charge: $\qquad E = \dfrac{Q}{4\pi\varepsilon_0 r^2}$

Electrostatic field due to line charge: $\qquad E = \dfrac{\lambda}{2\pi\varepsilon_0 r}$

Gravitational field due to point mass: $\qquad g = \dfrac{GM}{r^2}$

Magnetic field due to straight current-carrying wire: $\qquad B = \dfrac{\mu_0 I}{2\pi r}$

Particular expressions for potential

Due to a point charge: $\qquad \phi = \dfrac{Q}{4\pi\varepsilon_0 r}$

Due to a point mass: $\qquad \phi = -\dfrac{GM}{r}$

Capacitors

Capacitance of a parallel plate capacitor: $\qquad C = \varepsilon_0 A / d$

Energy stored in electric field per unit volume: $\qquad = \tfrac{1}{2}\varepsilon_0 E^2$

8.3 Rotation

Angular speed (ω), speed in circular motion (v), and magnitude of centripetal acceleration (a)

$$\omega = \frac{2\pi}{T}$$

$$v = \frac{2\pi}{T} r = \omega r$$

$$a = \omega v = \omega^2 r$$

Rotated coordinate system

$$\begin{pmatrix} x' \\ y' \\ z' \end{pmatrix} = \begin{pmatrix} \cos\theta & \sin\theta & 0 \\ -\sin\theta & \cos\theta & 0 \\ 0 & 0 & 1 \end{pmatrix} \begin{pmatrix} x \\ y \\ z \end{pmatrix}$$

$$\begin{pmatrix} x \\ y \\ z \end{pmatrix} = \begin{pmatrix} \cos\theta & -\sin\theta & 0 \\ \sin\theta & \cos\theta & 0 \\ 0 & 0 & 1 \end{pmatrix} \begin{pmatrix} x' \\ y' \\ z' \end{pmatrix}$$

Rotating vectors and the vector product

$$\hat{\mathbf{x}} \times \hat{\mathbf{x}} = \hat{\mathbf{y}} \times \hat{\mathbf{y}} = \hat{\mathbf{z}} \times \hat{\mathbf{z}} = 0$$
$$\hat{\mathbf{x}} \times \hat{\mathbf{y}} = -\hat{\mathbf{y}} \times \hat{\mathbf{x}} = \hat{\mathbf{z}}$$
$$\hat{\mathbf{y}} \times \hat{\mathbf{z}} = -\hat{\mathbf{z}} \times \hat{\mathbf{y}} = \hat{\mathbf{x}}$$
$$\hat{\mathbf{z}} \times \hat{\mathbf{x}} = -\hat{\mathbf{x}} \times \hat{\mathbf{z}} = \hat{\mathbf{y}}$$
$$\frac{d_a}{dt}\mathbf{r} = \frac{d_r}{dt}\mathbf{r} + \boldsymbol{\omega} \times \mathbf{r}$$

Angular velocity ($\boldsymbol{\omega}$), velocity ($\mathbf{v}$), and radius vector ($\mathbf{r}$)

$$\mathbf{v} = \boldsymbol{\omega} \times \mathbf{r}$$

Vector triple product

$$\mathbf{a} \times (\mathbf{b} \times \mathbf{c}) = \hat{\mathbf{x}}\{b_x(a_x c_x + a_y c_y + a_z c_z) - c_x(a_x b_x + a_y b_y + a_z b_z)\}$$
$$+ \hat{\mathbf{y}}\{b_y(a_x c_x + a_y c_y + a_z c_z) - c_y(a_x b_x + a_y b_y + a_z b_z)\}$$
$$+ \hat{\mathbf{z}}\{b_z(a_x c_x + a_y c_y + a_z c_z) - c_z(a_x b_x + a_y b_y + a_z b_z)\}$$
$$\mathbf{a} \times (\mathbf{b} \times \mathbf{c}) = (\mathbf{a} \cdot \mathbf{c})\mathbf{b} - (\mathbf{a} \cdot \mathbf{b})\mathbf{c}$$
$$(\mathbf{a} \times \mathbf{b}) \times \mathbf{c} = (\mathbf{c} \cdot \mathbf{a})\mathbf{b} - (\mathbf{c} \cdot \mathbf{b})\mathbf{a}$$

Acceleration vectors in rotating frames

$$\frac{d_a}{dt}\mathbf{v} = \frac{d_r}{dt}\mathbf{v} + \boldsymbol{\omega} \times \mathbf{v}$$
$$\frac{d_a}{dt}\mathbf{v} = \frac{d_r^2}{dt^2}\mathbf{r} + \frac{d_r}{dt}\boldsymbol{\omega} \times \mathbf{r} + 2\boldsymbol{\omega} \times \frac{d_r}{dt}\mathbf{r} + \boldsymbol{\omega} \times (\boldsymbol{\omega} \times \mathbf{r})$$

Cartesian ellipse and relationship between semi-major axis (a), semi-minor axis (b), and eccentricity (e)

$$\frac{x^2}{a^2} + \frac{y^2}{b^2} = 1, \quad a = \frac{b}{\sqrt{1-e^2}}$$

Polar coordinates (r, θ) for conic sections with d the directrix and e the eccentricity

$$r(\theta) = \frac{ed}{1 + e\cos\theta}$$

Equations of motion for a two-body planetary system

$$\mathbf{a} + G\frac{(m_1 + m_2)}{r^3}\mathbf{r} = \mathbf{0}$$

$$\mathbf{r} \times \mathbf{a} + \frac{G(m_1 + m_2)}{r^3}\mathbf{r} \times \mathbf{r} = \mathbf{0} \Rightarrow \frac{d}{dt}(\mathbf{r} \times \mathbf{v}) = \mathbf{0}$$

$$\mathbf{v} \cdot \mathbf{a} + \frac{G(m_1 + m_2)}{r^3}\mathbf{v} \cdot \mathbf{r} = \frac{d}{dt}\left\{\frac{v^2}{2} - \frac{G(m_1 + m_2)}{r}\right\} = 0$$

Constants of the motion ($\mathbf{h}$ and E) with μ the reduced mass

$$\mathbf{r} \times \mathbf{v} = \mathbf{h}$$

$$\frac{v^2}{2} - \frac{G(m_1 + m_2)}{r} = \frac{E}{\mu}$$

Kepler's laws

$$r(\theta) = \frac{h^2/G(m_1+m_2)}{(1 + e\cos\theta)}$$

$$\frac{d}{dt}\mathbf{A} = \frac{1}{2}h\hat{\mathbf{z}}$$

$$T^2 = \frac{4\pi^2}{G(m_1 + m_2)}a^3$$

8.4 Waves

Complex numbers

A complex number $z = x + iy$ contains a real part $\text{Re}(z) = x$, and an imaginary part $\text{Im}(z) = y$.

Basic arithmetic:
$$(a + ib) + (x + iy) = a + x + i(b + y)$$
$$(a + ib) - (x + iy) = a - x + i(b - y)$$
$$(a + ib)(x + iy) = ax - by + i(bx + ay)$$

The complex conjugate (z^*) of a complex number $z = x + iy$ is given by $z^* = x - iy$. The modulus of $z = x + iy$ is written $|z| = \sqrt{z^*z} = \sqrt{x^2 + y^2}$.

168 Summary of equations

The argument of $z = x + iy$ is written $\text{Arg}(z) = \tan^{-1}\left(\dfrac{y}{x}\right)$.

The complex number whose modulus is r and whose argument is θ can be written

$$re^{i\theta} = r\cos\theta + ir\sin\theta$$

It follows that

$$re^{i\theta} \times se^{i\phi} = rs\,e^{i(\theta+\phi)}$$
$$e^{i\theta} + e^{-i\theta} = 2\cos\theta$$
$$e^{i\theta} - e^{-i\theta} = 2i\sin\theta$$

Describing a simple harmonic oscillation

Angular frequency: $\quad \omega = 2\pi f$

Equation for displacement: $\quad y = A\cos(\omega t + \phi)$
$\quad\quad\quad\quad\quad\quad\quad\quad\quad\quad y = \text{Re}\,Ae^{i\omega t}$

Newton's second law: $\quad \dfrac{d^2 y}{dt^2} = -\omega^2 y$, $\quad y$ measured from equilibrium.

Describing a damped harmonic oscillation

Newton's second law: $\quad m\dfrac{d^2 y}{dt^2} + r\dfrac{dy}{dt} + ky = 0$

Solution: $\quad y = \text{Re}\left(e^{-rt/2m}\left(Ae^{i\omega t} + Be^{-i\omega t}\right)\right)$,

$$\text{where } \omega = \sqrt{\dfrac{k}{m} - \dfrac{r^2}{4m^2}}.$$

Waves in one dimension

Equation for displacement: $\quad y = \text{Re}\,Ae^{i(\omega t - kx)}$

Wavenumber: $\quad k = \dfrac{2\pi}{\lambda} = \dfrac{\omega}{c}$

Wave equation: $\quad \dfrac{\partial^2 y}{\partial x^2} = \dfrac{1}{c^2}\dfrac{\partial^2 y}{\partial t^2}$

For a wave on a string: $\quad c = \sqrt{\dfrac{T}{\rho}}$

Specific impedance: $\quad Z = \rho c = \dfrac{T}{c} = \dfrac{Tk}{\omega}$

Power transmitted: $\quad P = \dfrac{1}{2}ZA^2\omega^2$

Impedance matching

$$A_r = A_i \frac{Z_L - Z_R}{Z_L - Z_R}$$

$$A_t = A_i \frac{2Z_L}{Z_L - Z_R}$$

Waves in three dimensions

Plane wave: $\quad \mathbf{E} = \mathbf{E}_0 \cos(\omega t - \mathbf{k} \cdot \mathbf{r})$

Spherical wave: $\quad E = \dfrac{E_0}{r} \cos(\omega t - kr)$

8.5 Circuits

Direct current

Flow equation: $\quad I = quAn$

Alternating currents and voltages

Written as:
$$V = \mathrm{Re} V_0 e^{i\omega t}$$
$$I = \mathrm{Re} J_0 e^{i\omega t},$$

where complex current: $J_0 = I_0 e^{i\phi}$

Resistor: $\quad V = IR$

Capacitor: $\quad I = C \dfrac{dV}{dt}$

Inductor: $\quad V = L \dfrac{dI}{dt}$

Average power dissipated in resistor: $\quad P = V_{\mathrm{rms}} I_{\mathrm{rms}}$

$$P = I_{\mathrm{rms}}^2 R$$

$$V_{\mathrm{rms}} = \frac{V_0}{\sqrt{2}}, \quad I_{\mathrm{rms}} = \frac{I_0}{\sqrt{2}}$$

Average power dissipated in a circuit: $\quad P = V_{\mathrm{rms}} I_{\mathrm{rms}} \cos \phi$

Impedance

Definition: $\quad Z = \dfrac{V_0}{J_0}$

Generally: $\quad Z = R + iX$

where X is reactance.

170 *Summary of equations*

Resistor: $\quad Z = R$

Capacitor: $\quad Z = -\dfrac{i}{\omega C}$

Inductor: $\quad Z = i\omega L$

Series circuit: $\quad Z = Z_1 + Z_2 + Z_3 + \cdots$

Parallel circuit: $\quad \dfrac{1}{Z} = \dfrac{1}{Z_1} + \dfrac{1}{Z_2} + \dfrac{1}{Z_3} + \cdots$

8.6 Thermal physics

First law

$$\Delta U = \Delta Q + \Delta W$$
$$\Delta U = \Delta Q - p\Delta V$$

Thermodynamic temperature

$$\left|\dfrac{\Delta Q_1}{\Delta Q_2}\right| = \dfrac{T_1}{T_2}$$

Efficiency of a reversible heat engine

$$\eta = \left|\dfrac{\Delta W}{\Delta Q_1}\right| = \dfrac{\Delta Q_1 - |\Delta Q_2|}{\Delta Q_1} = 1 - \dfrac{T_2}{T_1}$$

Reversible processes

$$\sum_{\text{Complete cycle}} \dfrac{\Delta Q}{T} = 0 \Rightarrow \oint \dfrac{dQ}{T} = 0$$

Re-statement of the first law for a reversible process

$$\Delta U = \Delta Q + \Delta W = T\Delta S - p\Delta V$$
$$\Delta Q = T\Delta S$$
$$\Delta W = -p\Delta V$$

Boltzmann law

Probability that a particle has energy $E \propto e^{-E/kT}$

Perfect gases

$$pV = nRT = NkT$$

$$C_V = \frac{dQ_V}{dT} = \frac{dU}{dT}$$

$$C_P = \frac{dQ_P}{dT} = \frac{dU}{dT} + p\frac{dV}{dT}$$

$$= C_V + p\frac{d}{dT}\left(\frac{RT}{p}\right) = C_V + R$$

Monatomic: $C_V = \frac{3}{2}R$ and $C_P = C_V + R = \frac{5}{2}R$

Diatomic: $C_V = \frac{5}{2}R$ and $C_P = C_V + R = \frac{7}{2}R$

$$\gamma = \frac{C_P}{C_V}$$

Isothermal: $pV = \text{constant}$

Adiabatic: $pV^\gamma = \text{constant}$

Adiabatic: $\dfrac{T_1}{T_2}\dfrac{V_1^{\gamma-1}}{V_2^{\gamma-1}} = 1$

Workshop solutions

Chapter 1

1.1.3 Simple differential equations

(a) The preamble of the workshop solves the differential equation: $(dv/dt) = -kv$ and gives the solution: $v(t) = v_0 e^{-kt}$. It is therefore a simple matter to obtain an expression for $a(t)$:

$$a(t) = \frac{dv}{dt} = -kv = -kv_0 e^{-kt}.$$

We can obtain $s(t)$ by remembering that $(ds/dt) = v(t)$ and that the approximation $\Delta s = (ds/dt)\Delta t$:

$$\Delta s = v\Delta t = v_0 e^{-kt} \Delta t$$

gets better and better as $\Delta t \to 0$. The summation of all the little bits Δs becomes an integral in the limit $\Delta t \to 0$:

$$s(t) = \int_0^t v_0 e^{-kt'} dt' = \left[\frac{v_0 e^{-kt'}}{-k}\right]_0^t = \frac{v_0}{k}(1 - e^{-kt}).$$

(b)
$$\frac{dv}{dt} = g - kv$$

so,

$$\Delta v = \frac{dv}{dt}\Delta t = (g - kv)\Delta t \Rightarrow \frac{\Delta v}{(g - kv)} = \Delta t.$$

Once again, the sum of all the Δt becomes an integral in the limit $\Delta t \to 0$:

$$\int_0^v \frac{dv'}{(g - kv')} = \int_0^t dt',$$

which gives:

$$\left[-\frac{\ln(g - kv')}{k}\right]_0^v = \left[-\frac{\ln(g - kv)}{k} + \frac{\ln(g)}{k}\right] = t,$$

which simplifies to:
$$v(t) = \frac{g}{k}(1 - e^{-kt}).$$

Indeed this expression has the correct asymptotic value, as $t \to \infty$, $v \to g/k$, which is exactly what one would expect as $(dv/dt) \to 0$ in $(dv/dt) = g - kv$. Once again $a(t)$ is easily obtained:

$$a(t) = \frac{dv}{dt} = g - kv = g - k\left(\frac{g}{k}(1 - e^{-kt})\right) = ge^{-kt}.$$

$s(t)$ is obtained by following the same procedure as before, so

$$\Delta s = v \Delta t = \frac{g}{k}(1 - e^{-kt})\Delta t,$$

by integrating:

$$s(t) = \frac{g}{k}\int_0^t (1 - e^{-kt'})dt' = \frac{g}{k}\left[t' + \frac{e^{-kt'}}{k}\right]_0^t = \frac{g}{k}\left(t + \frac{1}{k}(e^{-kt} - 1)\right).$$

(c)
$$\frac{dv}{dt} = g - kv^2$$

so,

$$\Delta v = \frac{dv}{dt}\Delta t = (g - kv^2)\Delta t \Rightarrow \frac{\Delta v}{(g - kv^2)} = \Delta t,$$

which leads to the integral:

$$\int_0^v \frac{dv'}{\left(\frac{g}{k} - v'^2\right)} = k\int_0^t dt'.$$

By partial fractions:

$$\frac{1}{\left(\frac{g}{k} - v'^2\right)} = \frac{\frac{1}{2}\sqrt{\frac{k}{g}}}{\sqrt{\frac{g}{k}} - v'} + \frac{\frac{1}{2}\sqrt{\frac{k}{g}}}{\sqrt{\frac{g}{k}} + v'},$$

174 Workshop solutions

so the integral becomes

$$\frac{1}{2}\sqrt{\frac{k}{g}}\int_0^v \frac{dv'}{\left(\sqrt{\frac{g}{k}} - v'\right)} + \frac{1}{2}\sqrt{\frac{k}{g}}\int_0^v \frac{dv'}{\left(\sqrt{\frac{g}{k}} + v'\right)} = kt.$$

Performing the integration and putting in the limits we get

$$\left[-\ln\left(\sqrt{\frac{g}{k}} - v'\right)\right]_0^v + \left[\ln\left(\sqrt{\frac{g}{k}} + v'\right)\right]_0^v = \ln\left(\frac{\left(\sqrt{\frac{g}{k}} + v\right)}{\left(\sqrt{\frac{g}{k}} - v\right)}\right) = 2t\sqrt{gk},$$

which means that

$$\left(\frac{\left(\sqrt{\frac{g}{k}} - v\right)}{\left(\sqrt{\frac{g}{k}} + v\right)}\right) = e^{-2\sqrt{gk}\,t},$$

which after some rearrangement becomes:

$$v(t) = \sqrt{\frac{g}{k}}\left(\frac{e^{\sqrt{gk}\,t} - e^{-\sqrt{gk}\,t}}{e^{\sqrt{gk}\,t} + e^{-\sqrt{gk}\,t}}\right).$$

Now it just so happens that the combinations of exponential functions in the parentheses are in fact called hyperbolic functions:

$$\sinh(x) = \frac{e^x - e^{-x}}{2}$$
$$\cosh(x) = \frac{e^x + e^{-x}}{2}$$
$$\tanh(x) = \frac{\sinh(x)}{\cosh(x)} = \frac{e^x - e^{-x}}{e^x + e^{-x}}$$

so,

$$v(t) = \sqrt{\frac{g}{k}}\tanh\left(\sqrt{gk}\,t\right).$$

$a(t)$ is easily obtained either by differentiating or by using $(dv/dt) = g - kv^2$,

$$\frac{dv}{dt} = g - kv^2 = g - g\tanh^2\left(\sqrt{gk}\,t\right) = \frac{g}{\cosh^2\left(\sqrt{gk}\,t\right)}.$$

$s(t)$ is, as usual, obtained by integration:

$$s(t) = \int_0^t v(t')dt' = \sqrt{\frac{g}{k}} \int_0^t \tanh\left(\sqrt{gk}\,t'\right) dt' = \frac{1}{k}\left[\ln\left(\cosh\left(\sqrt{gk}\,t'\right)\right)\right]_0^t$$
$$= \frac{1}{k}\ln\left(\cosh\left(\sqrt{gk}\,t\right)\right).$$

1.1.5 Motion on the surface of a smooth inclined plane

(a) It is useful to have at our disposal the sine and cosine of the angles that are important to us in this problem:

$$\sin(30°) = \cos(60°) = \frac{1}{2}$$
$$\sin(60°) = \cos(30°) = \frac{\sqrt{3}}{2}.$$

Using the coordinate system suggested by the question, we have that the acceleration on the slope is

$$\mathbf{a} = \begin{pmatrix} 0 \\ -g\sin(30°) \end{pmatrix} = -\frac{1}{2}\begin{pmatrix} 0 \\ g \end{pmatrix}.$$

The velocity vector then in the suggested coordinate system is

$$\mathbf{v}(t) = \begin{pmatrix} v_x \\ v_y \end{pmatrix} = \begin{pmatrix} v\cos(60°) \\ v\cos(30°) - \frac{g}{2}t \end{pmatrix} = \frac{1}{2}\begin{pmatrix} v \\ \sqrt{3}v - gt \end{pmatrix}.$$

So $\mathbf{r}(t)$ is just:

$$\mathbf{r}(t) = \frac{1}{2}(\mathbf{v}(0+t) + \mathbf{v}(0))t = \frac{1}{2}\begin{pmatrix} vt \\ \sqrt{3}vt - \frac{g}{2}t^2 \end{pmatrix}.$$

For a given position (x, y), the x-component tells us that $t = 2x/v$, so

$$y = \frac{\sqrt{3}}{2}v\left(\frac{2x}{v}\right) - \frac{g}{4}\left(\frac{2x}{v}\right)^2 = \sqrt{3}x - \frac{gx^2}{v^2}.$$

(b) The trajectory described above is symmetric in the suggested coordinate system, so the maximum y-value will occur at the x midpoint. The extremities of the x-values are obtained by setting $y = 0$:

$$0 = x\left(\sqrt{3} - \frac{gx}{v^2}\right) \Rightarrow x = 0, \frac{\sqrt{3}v^2}{g}.$$

Midpoint occurs for $x = \sqrt{3}v^2/2g$, so maximum $y = \frac{3v^2}{4g}$.

176 Workshop solutions

1.2.4 The conservation of linear momentum

The vector equation:
$$\mathbf{P}_{i1} + \mathbf{P}_{i2} = \mathbf{P}_{f1} + \mathbf{P}_{f2} = \mathbf{P}$$

is in fact a shorthand for the two sets of equations:

$$\begin{pmatrix} m_1 u_1 \cos\theta \\ m_1 u_1 \sin\theta \end{pmatrix} + \begin{pmatrix} m_2 u_2 \cos\varphi \\ m_2 u_2 \sin\varphi \end{pmatrix} = \begin{pmatrix} m_1 v_1 \\ m_1 u_1 \sin\theta \end{pmatrix} + \begin{pmatrix} m_2 v_2 \\ m_2 u_2 \sin\varphi \end{pmatrix};$$

that is,

$$m_1 u_1 \cos\theta + m_2 u_2 \cos\varphi = m_1 v_1 + m_2 v_2$$
$$m_1 u_1 \sin\theta + m_2 u_2 \sin\varphi = \text{constant}.$$

Newton's law may be written mathematically as

$$\frac{v_1 - v_2}{u_1 \cos\theta - u_2 \cos\varphi} = -e,$$

To see this let us just go through the statement again:

When two bodies of given substances collide the relative velocity after impact $(v_1 - v_2)$ is in a constant ratio (e) to the relative velocity before impact, and in the opposite direction (hence the minus sign). If the bodies impinge obliquely the empirical law holds for the component velocities along the common normal $(u_1 \cos\theta - u_2 \cos\varphi)$.

(a) Obvious! Just take the statement and turn it into mathematics as is done above. The minus sign is taken into the denominator and we get

$$\frac{v_1 - v_2}{u_2 \cos\varphi - u_1 \cos\theta} = e.$$

(b) Using the answer for (a) we can find expressions for v_1 and v_2 and put them into the expression:

$$m_1 u_1 \cos\theta + m_2 u_2 \cos\varphi = m_1 v_1 + m_2 v_2$$

after some algebra:

$$v_1 = \frac{(m_1 - em_2)u_1 \cos\theta + m_2 u_2 (1+e)\cos\phi}{m_1 + m_2}$$

$$v_2 = \frac{(m_2 - em_1)u_2 \cos\phi + m_1 u_1 (1+e)\cos\theta}{m_1 + m_2}.$$

(c) With $m_1 = m_2$ and $u_2 = 0$

$$v_1 = \frac{(1-e)u_1 \cos\theta}{2}$$

$$v_2 = \frac{(1+e)u_1 \cos\theta}{2},$$

so v_1/v_2 is
$$\frac{v_1}{v_2} = \frac{(1-e)}{(1+e)}.$$

(d) When $e = 1$ we have what is called an *elastic collision*. The equations would suggest that $v_1 = 0$ so all the horizontal momentum particle (1) had is transferred to particle (2). Since particle (2) was initially at rest, $u_2 = 0$, the two particles will collide and move off in directions that are perpendicular to each other. The velocities must therefore be related through the Pythagoras theorem*:
$$u_1^2 = u_1^2 \sin^2 \theta + v_2^2 = u_1^2 \sin^2 \theta + u_1^2 \cos^2 \theta,$$
which, if you multiply by the mass m of the particles, is just
$$\frac{1}{2}mu_1^2 = \frac{1}{2}mu_1^2 \sin^2 \theta + \frac{1}{2}mu_1^2 \cos^2 \theta.$$

It is why we very often say that in elastic collisions the kinetic energy is conserved. With $e = 1$ we have the simple situation that the energy is not distributed amongst other types of energy in the collision so, along the common normal, the magnitude of the velocity of approach is equal to the magnitude of the velocity after restitution. Notice that when $e = 0$ we have that $v_1 = v_2$; that is, the two particles do not separate after impact (an *inelastic collision*).

1.2.6 Newton and the apple

(a) If the distance the Moon falls in 1 s is s' then
$$s' = \frac{1}{2}g't^2, \quad \text{with } g' = \frac{g}{3600},$$
so the distance fallen by the Moon in 1 s is essentially the distance an apple near the surface of the Earth falls in 1 s divided by $3600(60 \times 60)$, which gives $s' \cong 1.3$ mm.

(b) $v = \dfrac{2\pi R}{T} \cong 1$ km/s.

(c) If $(OB) = (R + \delta)$, then δ is the small distance that the Moon falls in 1 s. So
$$(OA)^2 + (AB)^2 = (OB)^2,$$
$$R^2 + (AB)^2 = (R+\delta)^2 = (R^2 + 2R\delta + \delta^2).$$
For small δ we can neglect terms of second order (i.e. δ^2) so
$$\delta \cong (AB)^2/2R,$$
or about 1.3 mm. The Moon does fall towards the Earth just enough each second to remain in its orbital motion.

---
*Of course you will remember that $\sin^2 \theta + \cos^2 \theta = 1$.

Chapter 2

2.2 Motion in a uniform field in 1-D

(a) From $v = gt$, it follows that $t = v/g$ so $y = \frac{1}{2}gt^2 = \frac{1}{2}g(v/g)^2 = v^2/2g$.

(b) Initial speed $= 0$; final speed $= v$; average speed $= \frac{1}{2}(0 + v) = \frac{1}{2}v$
Distance fallen $y =$ Average speed $\times$ Time $= \frac{1}{2}vt$
From (a), $t = v/g$, so $y = \frac{1}{2}vt = v^2/2g$, in agreement with (a).

(c) Using methods of part (b):
Initial speed $= u$; final speed $= v$; average speed $= \frac{1}{2}(u + v)$
From definition of acceleration $g = (v - u)/t$ so $t = (v - u)/g$
Distance fallen $=$ Average speed $\times$ Time

$$y = \frac{u+v}{2} \times \frac{v-u}{g} = \frac{v^2 - u^2}{2g}.$$

If you choose to use methods of part (a), you must start with $y = ut + \frac{1}{2}gt^2$.

(d) From (c) we have

$$mgy = mg\frac{u+v}{2} \times \frac{v-u}{g} = m\frac{v^2-u^2}{2} = \frac{1}{2}mv^2 - \frac{1}{2}mu^2.$$

(e) $y = h_1 - h_2$, so $mgy = mgh_1 - mgh_2$ and from (d)

$$mgy = mgh_1 - mgh_2 = \frac{1}{2}mv^2 - \frac{1}{2}mu^2$$

$$mgh_1 + \frac{1}{2}mu^2 = mgh_2 + \frac{1}{2}mv^2.$$

2.3 Scalar product of vectors

(a) Magnitude of **b** is b. Component of **a** parallel to **b** is $a\cos\theta$.
Multiplying them gives $\mathbf{a}\cdot\mathbf{b} = ba\cos\theta = ab\cos\theta$ as before.

(b) We visualize the vectors using Figure 2.3, with **a** aligned with the x-axis.
Then $\mathbf{a}\cdot\mathbf{b} = ab\cos\theta$. But $a = a_x$ and $b\cos\theta = b_x$ so $\mathbf{a}\cdot\mathbf{b} = a_x b_x$.

(c) We repeat the methods of (b), now aligning **a** with the y-axis.
Then $\mathbf{a}\cdot\mathbf{b} = ab\cos\theta$. But $a = a_y$ and $b\cos\theta = b_y$ so $\mathbf{a}\cdot\mathbf{b} = a_y b_y$.

(d) (i) $\mathbf{a}\cdot\mathbf{b} = ab\cos\beta$
(ii) $\mathbf{a}\cdot\mathbf{c} = ac\cos\gamma$
(iii) $\mathbf{a}\cdot\mathbf{d} = ad\cos\delta$
(iv) $\mathbf{a}\cdot\mathbf{b} + \mathbf{a}\cdot\mathbf{c} = a(b\cos\beta + c\cos\gamma)$
(v) From Figure 2.4, $b\cos\beta + c\cos\gamma = d\cos\delta$
Thus $\mathbf{a}\cdot\mathbf{b} + \mathbf{a}\cdot\mathbf{c} = ad\cos\delta = \mathbf{a}\cdot\mathbf{d} = \mathbf{a}\cdot(\mathbf{b}+\mathbf{c})$.

(e)
$$\begin{pmatrix}a_x\\a_y\end{pmatrix}\cdot\begin{pmatrix}b_x\\b_y\end{pmatrix}=\left\{\begin{pmatrix}a_x\\0\end{pmatrix}+\begin{pmatrix}0\\a_y\end{pmatrix}\right\}\cdot\begin{pmatrix}b_x\\b_y\end{pmatrix}$$
$$=\begin{pmatrix}a_x\\0\end{pmatrix}\cdot\begin{pmatrix}b_x\\b_y\end{pmatrix}+\begin{pmatrix}0\\a_y\end{pmatrix}\cdot\begin{pmatrix}b_x\\b_y\end{pmatrix}$$
$$=a_xb_x+a_yb_y$$

2.4 Motion in a uniform field in 3-D

(a)
$$\mathbf{g}\cdot\mathbf{s}=\frac{\mathbf{v}-\mathbf{u}}{t}\cdot\frac{1}{2}(\mathbf{u}+\mathbf{v})t$$
$$=\frac{1}{2}(\mathbf{v}-\mathbf{u})\cdot(\mathbf{u}+\mathbf{v})$$
$$=\frac{1}{2}(\mathbf{v}\cdot\mathbf{u}-\mathbf{u}\cdot\mathbf{u}-\mathbf{u}\cdot\mathbf{v}+\mathbf{v}\cdot\mathbf{v}).$$

But we showed in 2.3 (a) that $\mathbf{a}\cdot\mathbf{b}=\mathbf{b}\cdot\mathbf{a}$, so this means that $\mathbf{u}\cdot\mathbf{v}=\mathbf{v}\cdot\mathbf{u}$. We also noted that $\mathbf{a}\cdot\mathbf{a}=a^2$, and hence $\mathbf{u}\cdot\mathbf{u}=u^2$. It follows that

$$\mathbf{g}\cdot\mathbf{s}=\frac{1}{2}v^2-\frac{1}{2}u^2.$$

(b) Work done $=\mathbf{F}\cdot\mathbf{s}=m\mathbf{g}\cdot\mathbf{s}=\frac{1}{2}mv^2-\frac{1}{2}mu^2=$ Gain in kinetic energy.
(c) Work done $=\mathbf{F}\cdot\mathbf{s}=q\mathbf{E}\cdot\mathbf{s}=$ Loss in potential energy. (Note that this must be a loss since the kinetic energy has increased.)
Change in potential energy $=-q\mathbf{E}\cdot\mathbf{s}$.
Change in potential (that is energy per unit charge) $=-\mathbf{E}\cdot\mathbf{s}$.
(d) Let us take a straight route $\mathbf{s}$ from a point on the negative plate to one on the positive plate along a line which makes angle θ to the perpendicular distance d. The vector $\mathbf{E}$ points from $+$ to $-$ (opposite to our direction of motion) perpendicular to the plates, and hence the change in potential is $-\mathbf{E}\cdot\mathbf{s}=Es\cos\theta=Ed$. This is independent of the angle θ or the length of the route s and thus all points on the positive plate must be at the same potential with respect to our chosen point on the negative plate. By a reverse argument from a fixed point on the positive plate to any point on the negative plate, we can also show that all points on the positive plate are at the same potential with respect to *any* point on the negative plate.

2.6 Evaluating line integrals

(a) By symmetry, the magnetic field strength must be the same at all points with the same distance r from the wire. The field lines take the form of a circle round the wire, and if we follow one for one circuit of the wire, we will travel a distance $2\pi r$ and the magnetic field will be the same all the way round (since we have not got closer to or further away from the wire). Given that we are following the wire, $\mathbf{B}$ is parallel to $\delta\mathbf{s}$, and therefore the $\cos\theta$ term equals 1.

Thus $\oint \mathbf{B} \cdot d\mathbf{s} = BS = 2\pi r B$, and since we know that this is equal to $\mu_0 I$, it follows that $B = (\mu_0 I)/(2\pi r)$.

(b)
$$\Delta\phi = -\int \mathbf{E} \cdot d\mathbf{s} = -\int E\, ds = -\int \frac{Q}{4\pi\varepsilon_0 r^2} dr = \left[\frac{Q}{4\pi\varepsilon_0 r}\right]_R^{2R} = -\frac{Q}{8\pi\varepsilon_0 R}$$

The first stage follows since our path $\mathbf{s}$ points away from the charge Q and is therefore in the same direction as $\mathbf{E}$. The next stage follows since $\delta s = \delta r$.

(c) For our path, $x = y = z$ at all points. With this simplification the electric field along our path becomes $\mathbf{E} = \frac{F}{a^2}\begin{pmatrix} x^2 \\ x^2 \\ x^2 \end{pmatrix}$, and so $\int \mathbf{E} \cdot d\mathbf{s}$ becomes $\int_0^1 E_x dx + \int_0^1 E_y dy + \int_0^1 E_z dz = 3\int_0^1 E_x dx$ since along our path $E_x = E_y = E_z$ and $\delta x = \delta y = \delta z$. We then evaluate this as

$$\Delta\phi = -\int \mathbf{E} \cdot d\mathbf{s} = -3\int_0^1 \frac{Fx^2}{a^2} dx = -3\left[\frac{Fx^3}{3a^2}\right]_0^1 = -\frac{F}{a^2}.$$

(d) Stage one: $\Delta\phi = -\int \mathbf{E} \cdot d\mathbf{s} = -\int E_x dx = -\int_0^1 ((Fyz)/a^2) dx = 0$ since y and z are both 0.
Stage two: $\Delta\phi = -\int \mathbf{E} \cdot d\mathbf{s} = -\int E_y dy = -\int_0^1 ((Fxz)/a^2) dy = 0$ since $z = 0$.
Stage three: $\Delta\phi = -\int \mathbf{E} \cdot d\mathbf{s} = -\int E_z dz = -\int_0^1 ((Fxy)/a^2) dz = -\int_0^1 ((F)/a^2) dz$
$= -\left[\frac{Fz}{a^2}\right]_0^1 = -F/a^2$, where we remember that as we go from (1,1,0) to (1,1,1), $x = y = 1$.
Adding these gives the same result as in (c).

(e) (i) $(d/dt)v^2 = d/dt\, (\mathbf{v}\cdot\mathbf{v}) = d\mathbf{v}/dt \cdot \mathbf{v} + \mathbf{v} \cdot d\mathbf{v}/dt = 2\mathbf{v} \cdot d\mathbf{v}/dt = 2\mathbf{v} \cdot \mathbf{g}$
(ii) $-\int m\mathbf{g} \cdot \mathbf{v}\, dt = -\int \frac{1}{2}m\, (d(v^2)/dt)\, dt = -\frac{1}{2}m\int d(v^2)$

If we now perform this integral from initial speed U to final speed V we get

$$-\int m\mathbf{g}\cdot\mathbf{v}\,dt = -\frac{1}{2}m\int_{U^2}^{V^2} d(v^2) = -\frac{1}{2}m\,(V^2 - U^2) = \frac{1}{2}m\,(U^2 - V^2),$$

which is the loss in kinetic energy.

(f) (i) $m\mathbf{a} =$ Resultant force $= \mathbf{F} + m\mathbf{g}$, thus $\mathbf{F} = m\mathbf{a} - m\mathbf{g} = m(\mathbf{a} - \mathbf{g})$.
(ii) $\int \mathbf{F} \cdot d\mathbf{s} = \int m\,(\mathbf{a} - \mathbf{g}) \cdot d\mathbf{s} = \int m\mathbf{a} \cdot d\mathbf{s} - \int m\mathbf{g} \cdot d\mathbf{s}$
The second term is the gain in potential energy $m\Delta\phi$ as in equation (2.12).
(iii) In (e) we showed that $-\int m\mathbf{g} \cdot d\mathbf{s} = -\int m\mathbf{g} \cdot \mathbf{v}\,dt = -\int \frac{1}{2}m\,(d(v^2)/dt)\,dt = -\frac{1}{2}m\int d(v^2)$ was the loss in kinetic energy. In that part, $\mathbf{g}$ was the acceleration, since the gravitational force was the only force acting.

Accordingly, to use this result in part (f) we have to replace the **g** with an **a**. Thus $-\int m\mathbf{a}\cdot\mathbf{v}\,dt$ will be the loss in kinetic energy, and therefore $\int m\mathbf{a}\cdot\mathbf{v}\,dt$ will be the gain in kinetic energy, as required.

2.8.1 The electrostatic field surrounding a charged wire

(a) Field lines radiate outward from the wire, and only pass through the curved surface. This surface has area $2\pi rl$. So the number of field lines $n = E \times 2\pi rl$, so $E = n/2\pi rl$.

(b) The total charge enclosed is $l\lambda$, and so the number of field lines must also be $Q/\varepsilon_0 = l\lambda/\varepsilon_0$.

(c) Thus $2\pi rlE = l\lambda/\varepsilon_0$ and so $E = \lambda/2\pi\varepsilon_0 r$.

(d) $\lambda = 4 \times 10^{-7}$ C/m, $r = 0.02$ m, ε_0 is given in data on inside back cover so,

$$E = \frac{4.00 \times 10^{-7} \text{ C/m}}{2\pi \times 8.85 \times 10^{-12} \text{ F/m} \times 0.02 \text{ m}} = 3.60 \times 10^5 \text{ V/m}.$$

(e) Number of free electrons in 1 m of wire = volume × density ÷ mass of one atom $= \pi \times 10^{-6}$ m$^2 \times 1$ m $\times 8930$ kg/m$^3 \div 1.07 \times 10^{-27}$ kg $= 2.62 \times 10^{25}$ electrons.

Charge on the free electrons in 1 m of wire $= 1.6 \times 10^{-19}$ C $\times 2.62 \times 10^{25} = 4.20 \times 10^6$ C.

Field at a distance of 200 m given by

$$E = \frac{4.20 \times 10^6 \text{ C/m}}{2\pi \times 8.85 \times 10^{-12} \text{ F/m} \times 200 \text{ m}} = 3.78 \times 10^{14} \text{ V/m}.$$

This is ridiculously large, but remember that the free electrons are not the only charged objects in the wire – you also have the bound electrons and the nuclei which have equal and opposite charge to the electrons. In fact the fields due to the positive and negative charges do not quite cancel out because of relativistic effects, and this can be used to explain the mutual attraction of two parallel current-carrying wires. That is a topic beyond this book, however.

Chapter 3

3.1.2 Rotated coordinate systems and matrices

(a) & (b) From the Figure 3.4 we can see:

$$\begin{aligned} x' &= \cos\theta\, x & +\sin\theta\, y & +0z \\ y' &= -\sin\theta\, x & +\cos\theta\, y & +0z\,. \\ z' &= 0x & +0y & +1z \end{aligned}$$

Now matrix multiplication goes like this:

$$\begin{pmatrix} v'_1 \\ v'_2 \\ v'_3 \end{pmatrix} = \begin{pmatrix} M_{11} & M_{12} & M_{13} \\ \cdot & \cdot & \cdot \\ \cdot & \cdot & \cdot \end{pmatrix} \begin{pmatrix} v_1 \\ v_2 \\ v_3 \end{pmatrix} \Rightarrow \begin{matrix} v'_1 = M_{11}v_1 + M_{12}v_2 + M_{13}v_3 \\ \cdot \\ \cdot \end{matrix}$$

$$\begin{pmatrix} v'_1 \\ v'_2 \\ v'_3 \end{pmatrix} = \begin{pmatrix} \cdot & \cdot & \cdot \\ M_{21} & M_{22} & M_{23} \\ \cdot & \cdot & \cdot \end{pmatrix} \begin{pmatrix} v_1 \\ v_2 \\ v_3 \end{pmatrix} \Rightarrow \begin{matrix} \cdot \\ v'_2 = M_{21}v_1 + M_{22}v_2 + M_{23}v_3. \\ \cdot \end{matrix}$$

$$\begin{pmatrix} v'_1 \\ v'_2 \\ v'_3 \end{pmatrix} = \begin{pmatrix} \cdot & \cdot & \cdot \\ \cdot & \cdot & \cdot \\ M_{31} & M_{32} & M_{33} \end{pmatrix} \begin{pmatrix} v_1 \\ v_2 \\ v_3 \end{pmatrix} \Rightarrow \begin{matrix} \cdot \\ \cdot \\ v'_3 = M_{31}v_1 + M_{32}v_2 + M_{33}v_3. \end{matrix}$$

So our expressions relating (x', y', z') and (x, y, z) can be written in matrix form:

$$\begin{pmatrix} x' \\ y' \\ z' \end{pmatrix} = \begin{pmatrix} \cos\theta & \sin\theta & 0 \\ -\sin\theta & \cos\theta & 0 \\ 0 & 0 & 1 \end{pmatrix} \begin{pmatrix} x \\ y \\ z \end{pmatrix} = \underline{R}(\theta) \begin{pmatrix} x \\ y \\ z \end{pmatrix}.$$

The matrix $\underline{R}(\theta)$ rotates the coordinate axes by an angle θ in the direction indicated in the diagram and the new coordinates of a point (x, y, z) are given by (x', y', z') in the new *rotated* coordinate system.

Now it just so happens that another matrix:

$$\underline{R}(\theta)^T = \begin{pmatrix} \cos\theta & -\sin\theta & 0 \\ \sin\theta & \cos\theta & 0 \\ 0 & 0 & 1 \end{pmatrix},$$

which is easily obtained from the first by swapping the columns for rows and the rows for columns (this operation is called *transposing*), and it has a special relationship with the first matrix. We call this matrix the *transpose* of the first, $\underline{R}(\theta)^T$.

Before we go any further, let us first spend some time reminding ourselves about matrix multiplication of two matrices like the ones we have above, for example,

$$\begin{pmatrix} M_{11} & M_{12} & M_{13} \\ M_{21} & M_{22} & M_{23} \\ M_{31} & M_{32} & M_{33} \end{pmatrix} \times \begin{pmatrix} N_{11} & N_{12} & N_{13} \\ N_{21} & N_{22} & N_{23} \\ N_{31} & N_{32} & N_{33} \end{pmatrix}.$$

The rules of matrix multiplication say that

$$\begin{pmatrix} M_{11} & M_{12} & M_{13} \\ \cdot & \cdot & \cdot \\ \cdot & \cdot & \cdot \end{pmatrix} \times \begin{pmatrix} N_{11} & \cdot & \cdot \\ N_{21} & \cdot & \cdot \\ N_{31} & \cdot & \cdot \end{pmatrix} = \begin{pmatrix} (MN)_{11} & \cdot & \cdot \\ \cdot & \cdot & \cdot \\ \cdot & \cdot & \cdot \end{pmatrix};$$

that is,
$$(MN)_{11} = M_{11}N_{11} + M_{12}N_{21} + M_{13}N_{31},$$

which is the (1,1)th element of the matrix that is the product of the two matrices. Let us try another element:

$$\begin{pmatrix} . & . & . \\ M_{21} & M_{22} & M_{23} \\ . & . & . \end{pmatrix} \times \begin{pmatrix} . & N_{12} & . \\ . & N_{22} & . \\ . & N_{32} & . \end{pmatrix} = \begin{pmatrix} . & . & . \\ . & (MN)_{22} & . \\ . & . & . \end{pmatrix};$$

that is,
$$(MN)_{22} = M_{21}N_{12} + M_{22}N_{22} + M_{23}N_{32}.$$

In general,
$$(MN)_{ij} = \sum_k M_{ik}N_{kj},$$

which turns out to be true for any square matrices (here we are of course only dealing with 3×3 matrices). You may have noticed that the letter, or sometimes we call it index, that is being summed over (k in this case) is always repeated. Indeed, this was noticed by Albert Einstein who introduced a convenient convention where: 'any repeated index in an expression like the one above, must indicate a summation, hence we can drop the summation sign' and

$$(MN)_{ij} = M_{ik}N_{kj}.$$

We can take things further then and write down the elements of a triple product of matrices, thus,

$$(LMN)_{ij} = L_{ip}M_{pq}N_{qj},$$

where the summations are implied over p and q. One last thing, it is not difficult to see that in general $(MN)_{ij} \neq (NM)_{ij}$; that is, in general, the order of multiplication of matrices matters.

Let us return to our rotation matrices $\underline{R}(\theta)$ and $\underline{R}(\theta)^T$.

The product
$$\underline{R}(\theta)^T \underline{R}(\theta) = \underline{R}(\theta)\underline{R}(\theta)^T = \begin{pmatrix} 1 & 0 & 0 \\ 0 & 1 & 0 \\ 0 & 0 & 1 \end{pmatrix}.$$

Here we notice that in this particular case order of multiplication does not matter. The matrix on the far right is called the *unit* matrix – multiplying it into any matrix or vector always gives you back the same matrix or vector. If applying $\underline{R}(\theta)$ is to rotate the coordinate axes, then the result above suggests

184 *Workshop solutions*

that $\underline{R}(\theta)^T = \underline{R}(-\theta)$, or a rotation by the same angle in the *opposite* direction. Since this rotation undoes the first rotation we call these two rotations *inverses* of each other. Obtaining the transpose of a matrix is not always the procedure to obtain the inverse (you will need to go to a maths textbook to find out about the method to obtain the inverse of a general matrix). The transpose is the inverse for a special class of matrices called *orthogonal* matrices. Rotation matrices for 3-D space are always orthogonal.

Therefore,

$$\begin{pmatrix} x \\ y \\ z \end{pmatrix} = \underline{R}(-\theta) \begin{pmatrix} x' \\ y' \\ z' \end{pmatrix} = \begin{pmatrix} \cos\theta & -\sin\theta & 0 \\ \sin\theta & \cos\theta & 0 \\ 0 & 0 & 1 \end{pmatrix} \begin{pmatrix} x' \\ y' \\ z' \end{pmatrix},$$

rotating the primed coordinates back by θ must bring us back to the unprimed coordinates.

3.1.3 Rotating vectors and the vector product

In this workshop we use the matrices of the previous workshop to keep track of the effect of *rotating* our points of view. The matrices in the previous workshop allow us to relate the coordinates of a point (x, y, z) to the coordinates of the same point in a rotated coordinate system (x', y', z'). To obtain the effect of *rotating* rather than just being *rotated* we use the operation of differentiation with respect to time. This is because to predict how the *rotating* coordinates evolve in time, we need their rates of change.

Let us set the primed coordinate system rotating at a constant rate:

$$\omega = \frac{2\pi}{T},$$

where T is the time period of the rotation. So $\theta = \theta(t) = \omega t$. From the previous workshop we have

$$\begin{pmatrix} x \\ y \\ z \end{pmatrix} = \begin{pmatrix} \cos\omega t & -\sin\omega t & 0 \\ \sin\omega t & \cos\omega t & 0 \\ 0 & 0 & 1 \end{pmatrix} \begin{pmatrix} x' \\ y' \\ z' \end{pmatrix},$$

which actually means

$$\begin{aligned} x &= \cos\omega t\, x' & -\sin\omega t\, y' & \quad +0z' \\ y &= \sin\omega t\, x' & +\cos\omega t\, y' & \quad +0z'. \\ z &= 0x' & +0y' & \quad +1z' \end{aligned}$$

(a) We can differentiate these three equations to obtain:

$$\frac{dx}{dt} = -\omega \sin\omega t\, x' + \cos\omega t \frac{dx'}{dt} \quad -\omega\cos\omega t\, y' - \sin\omega t \frac{dy'}{dt} \quad +0$$

$$\frac{dy}{dt} = \omega\cos\omega t\, x' + \sin\omega t \frac{dx'}{dt} \quad -\omega\sin\omega t\, y' + \cos\omega t \frac{dy'}{dt} \quad +0 \;.$$

$$\frac{dz}{dt} = \quad\quad +0 \quad\quad +0 \quad\quad +\frac{dz'}{dt}$$

Let us gather like terms together:

$$\frac{dx}{dt} = \cos\omega t \frac{dx'}{dt} \quad -\sin\omega t \frac{dy'}{dt} \quad +0 \quad -\omega\sin\omega t\, x' \quad -\omega\cos\omega t\, y' \quad +0$$

$$\frac{dy}{dt} = \sin\omega t \frac{dx'}{dt} \quad +\cos\omega t \frac{dy'}{dt} \quad +0 \quad +\omega\cos\omega t\, x' \quad -\omega\sin\omega t\, y' \quad +0\;.$$

$$\frac{dz}{dt} = 0 \quad +0 \quad +1\frac{dz'}{dt} \quad +0 \quad +0 \quad +0$$

From this grouping the following pattern is obvious:

$$\begin{pmatrix}\dot{x}\\\dot{y}\\\dot{z}\end{pmatrix} = \begin{pmatrix}\cos\omega t & -\sin\omega t & 0\\ \sin\omega t & \cos\omega t & 0\\ 0 & 0 & 1\end{pmatrix}\begin{pmatrix}\dot{x}'\\\dot{y}'\\\dot{z}'\end{pmatrix} + \omega\begin{pmatrix}-\sin\omega t & -\cos\omega t & 0\\ \cos\omega t & -\sin\omega t & 0\\ 0 & 0 & 0\end{pmatrix}\begin{pmatrix}x'\\y'\\z'\end{pmatrix},$$

where we have used the 'dot' notation for time derivatives to save room.

(b) In this part we concentrate on the second matrix term, which is actually a vector whose:

$$\begin{aligned} x\text{-component} &= -\omega\sin\omega t\, x' - \omega\cos\omega t\, y' + 0z'\\ y\text{-component} &= \omega\cos\omega t\, x' - \omega\sin\omega t\, y' + 0z'\\ z\text{-component} &= 0x' + 0y' + 0z' \end{aligned}\;.$$

From our original relationships between (x, y, z) and (x', y', z'), we see that this means the vector's

$$\begin{aligned} x\text{-component} &= -\omega y\\ y\text{-component} &= \omega x \quad\text{or}\quad \omega\begin{pmatrix}-y\\ x\\ 0\end{pmatrix}\;.\\ z\text{-component} &= 0 \end{aligned}$$

(c) Using the rules for the *vector product* we see

$$\omega\hat{\mathbf{z}} \times (x\hat{\mathbf{x}} + y\hat{\mathbf{y}} + z\hat{\mathbf{z}}) = \omega x\hat{\mathbf{y}} - \omega y\hat{\mathbf{x}} = \omega\begin{pmatrix}-y\\ x\\ 0\end{pmatrix}\;.$$

3.1.5 Vector triple product

(a) You just have to work through this by doing all the possible multiplications in $(\mathbf{b} \times \mathbf{c}) = (b_x\hat{\mathbf{x}} + b_y\hat{\mathbf{y}} + b_z\hat{\mathbf{z}}) \times (c_x\hat{\mathbf{x}} + c_y\hat{\mathbf{y}} + c_z\hat{\mathbf{z}})$ taking care to respect the order

of each multiplication that involves unit vectors. You should discover when you collect all the like terms together that this product comes to
$$\hat{\mathbf{x}}(b_y c_z - b_z c_y) + \hat{\mathbf{y}}(b_z c_x - b_x c_z) + \hat{\mathbf{z}}(b_x c_y - b_y c_x).$$
Footnote 5 of chapter 3 that appears with this answer is for those of you who have studied *determinants*. It just so happens that the expression above can be written shorthand as a determinant – do not worry if you have not yet studied them, they appear in more advanced work and when you really need them you will probably receive maths courses on them.

(b) Just as in (a), you will need to work through this to see how it emerges – it will be good algebra practice. Once again, respect the order of multiplication with unit vectors.

(c)
$$\hat{\mathbf{x}}\{b_x(a_x c_x + a_y c_y + a_z c_z) - c_x(a_x b_x + a_y b_y + a_z b_z)\}$$
$$+ \hat{\mathbf{y}}\{b_y(a_x c_x + a_y c_y + a_z c_z) - c_y(a_x b_x + a_y b_y + a_z b_z)\}$$
$$+ \hat{\mathbf{z}}\{b_z(a_x c_x + a_y c_y + a_z c_z) - c_z(a_x b_x + a_y b_y + a_z b_z)\}.$$

In this expression, we can try to group together like terms:
$$(a_x c_x + a_y c_y + a_z c_z)(b_x \hat{\mathbf{x}} + b_y \hat{\mathbf{y}} + b_z \hat{\mathbf{z}})$$
$$- (a_x b_x + a_y b_y + a_z b_z)(c_x \hat{\mathbf{x}} + c_y \hat{\mathbf{y}} + c_z \hat{\mathbf{z}})$$

which is just:
$$(\mathbf{a} \cdot \mathbf{c})\mathbf{b} - (\mathbf{a} \cdot \mathbf{b})\mathbf{c}.$$

(d) We know that $(\mathbf{a} \times \mathbf{b}) \times \mathbf{c} = -\mathbf{c} \times (\mathbf{a} \times \mathbf{b}) = \mathbf{c} \times (\mathbf{b} \times \mathbf{a})$, so taking the last expression and swapping the a and c terms we get:
$$(a_x c_x + a_y c_y + a_z c_z)(b_x \hat{\mathbf{x}} + b_y \hat{\mathbf{y}} + b_z \hat{\mathbf{z}})$$
$$- (c_x b_x + c_y b_y + c_z b_z)(a_x \hat{\mathbf{x}} + a_y \hat{\mathbf{y}} + a_z \hat{\mathbf{z}}),$$

which is of course just:
$$(\mathbf{a} \cdot \mathbf{c})\mathbf{b} - (\mathbf{c} \cdot \mathbf{b})\mathbf{a}.$$

3.2.3 Kepler's second law

(a) The angle ABC is $180° - \phi$ and $\sin(180° - \phi) = \sin \phi$. Therefore $r\Delta r \sin \phi$ really is the area of the parallelogram ABCD (Figure A.1).

(b) The area of the triangle ABC is the area swept in the time Δt and is exactly half the area of the parallelogram ABCD, and since:
$$|r\delta||\Delta \mathbf{r}| \sin \phi \, \hat{z} = r\Delta r \sin \phi \, \hat{\mathbf{z}}$$
we must have that the area of the triangle is just:
$$\Delta \mathbf{A} = \frac{1}{2} \mathbf{r} \times \Delta \mathbf{r}.$$

(c) $\dfrac{d}{dt}\mathbf{A} = \dfrac{1}{2}\mathbf{r} \times \dfrac{d}{dt}\mathbf{r} = \dfrac{1}{2}\mathbf{r} \times \mathbf{v} = \dfrac{1}{2}\mathbf{h} = \dfrac{1}{2}h\hat{\mathbf{z}}.$

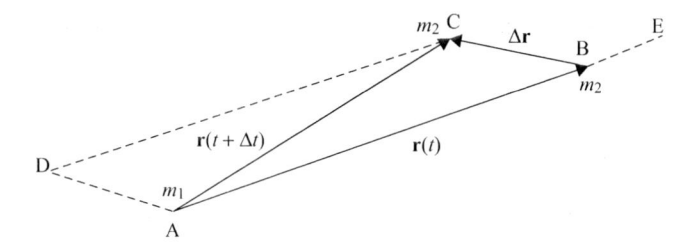

Fig. A.1

3.2.4 Kepler's third law

(a) The area enclosed in the ellipse is πab so as the rate at which area is swept is a constant h we must have that

$$h = \frac{\pi ab}{T}.$$

Comparing the two expressions:

$$r(\theta) = \frac{p}{1 + e\cos\theta}$$

and

$$r = \frac{h^2/G(m_1 + m_2)}{1 + e\cos(\theta)}$$

we see that

$$p = \frac{h^2}{G(m_1 + m_2)} \Rightarrow h = \sqrt{pG(m_1 + m_2)},$$

which of course means that

$$\frac{\pi ab}{T} = \sqrt{pG(m_1 + m_2)}.$$

When $\theta = 0, r(\theta) = a - ea$. So

$$r(0) = a(1-e) = \frac{p}{1+e} \Rightarrow p = a(1 - e^2).$$

(b)
$$\frac{\pi ab}{T} = \frac{\pi a^2\sqrt{1-e^2}}{T} = \sqrt{pG(m_1 + m_2)} = \sqrt{a(1-e^2)G(m_1 + m_2)}$$

$$\Rightarrow T^2 = \frac{4\pi^2}{G(m_1 + m_2)} a^3.$$

Chapter 4

4.1.1 Simple harmonic motion

(a) (i) $F = \dfrac{d(mu)}{dt} = m\dfrac{du}{dt} = m\dfrac{d^2y}{dt^2}$.

Now, here $F = -ky$, so $-ky = m(d^2y/dt^2)$, hence $(d^2y/dt^2) = -(k/m)y$. This is the same as $a = -\omega^2 y$ as in equation (4.9) providing that $\omega^2 = k/m$. Thus $\omega = \sqrt{(k/m)} = \sqrt{(1.2\ \text{N/m})/0.3\ \text{kg}} = 2.0$ rad/s.

(ii) If $y = C\cos\omega t + D\sin\omega t$, then $u = dy/dt = -C\omega\sin\omega t + D\omega\cos\omega t$. At $t=0$ we have $y = C$ and $u = D\omega$. Here $y = 2$ m and $u = 0$. Thus $C = 2$ m and $D = 0$. The equation of the motion is thus $y = 2$ m $\times \cos\omega t$, and the amplitude is 2 m since the largest value $\cos\omega t$ can take is 1.

(iii) Using the same logic as in (ii), at $t = 0$ we have $y = C = 0$ and $u = D\omega = 6$ m/s. Thus $D = 6$ m/s $\div 2$ rad/s $= 3$ m and $y = 3$ m $\times \sin\omega t$. The amplitude is 3 m.

(iv) At $t = 0$ we have $y = C = 2$ m and $u = D\omega = -1.5$ m/s. $D = u/\omega = -0.75$ m. Thus $y = 2$ m $\times \cos\omega t - 0.75$ m $\times \sin\omega t$.

Using the conversion formulae (4.8) we have $A = \sqrt{2^2 + 0.75^2}$ m $= 2.14$ m and $\phi = \tan^{-1}(0.75/2) = 20.6°$. Note that $\phi = 200.6°$ would also solve this equation, and so we have to check which answer is correct. Using equation (4.7) we note that if $\phi = 20.6°$ then both $\sin\phi$ and $\cos\phi$ would be positive, and hence C would be positive and D would be negative. This fits with our conditions, and hence the correct answer is $\phi = 20.6°$, or $20.6\pi/180$ rad $= 0.359$ rad. We use the answer in radians given that our ω is in rad/s.

So $y = 2.14$ m $\times \cos(\omega t + 0.359\ \text{rad})$.

The amplitude is accordingly 2.14 m.

$$u = dy/dt = -4.28\ \text{m/s} \times \sin(\omega t + 0.359\ \text{rad}).$$

This will first be zero when $\omega t + 0.359$ rad $= \pi$ rad and hence:

$$t = \dfrac{\pi\ \text{rad} - 0.359\ \text{rad}}{2\ \text{rad/s}} = 1.39\ \text{s}.$$

(b) (i) Since $F = -ky + h$, so $-ky + h = m(d^2y/dt^2)$, hence $(d^2y/dt^2) = -(k/m)y + (h/m)$.

Thus $P = -k/m$ and $Q = h/m$.

(ii) When you differentiate this expression for y, to find d^2y/dt^2, you will only ever have cosines or sines, never any constant part like the Q term.

(iii) At equilibrium $F = 0$ and so $-ky + h = 0$ and hence $y_0 = h/k$. Here this is at $y_0 = 0.375$ m.

(iv) Since $y = y' + y_0 = y' + h/k$, Newton's second law now takes the form $\left(d^2\left(y' + \tfrac{h}{k}\right)/dt^2\right) = -(k/m)(y' + h/k) + h/m = -(ky'/m)$. Now since h/k is a constant, we can write $(d^2(y' + (h/k))/dt^2) = (d^2(y')/dt^2)$, and accordingly, $(d^2y'/dt^2) = -(ky'/m)$, which is a regular simple harmonic motion equation in y' with $\omega^2 = k/m$ as before. Therefore y' can be written in the form $y' = A\cos(\omega t + \phi)$, and

$$y = \frac{h}{k} + A\cos(\omega t + \phi).$$

(c) No it will not, because the acceleration is not simply proportional to $-y$. However given that it is only knocked out of position by 3 cm, the contribution of the squared term will be negligible in comparison to the linear term. The largest value of the linear term $ky = 1.2$ N/m $\times$ 0.03 m $= 0.036$ N, whereas the largest value of the squared term is $hy^2 = 2$ N/m^2 $\times$ (0.03 m)^2 = 0.0018 N which is 20 times smaller. Thus, while it is not strictly correct, simple harmonic motion may be a suitable approximation to the motion. It all depends on the level of accuracy needed.

4.2 Describing complex numbers

(a) It moves one unit to the 'right' on Figure 4.2.
(b) It moves one unit upwards on Figure 4.2.
(c) Both its 'x-coordinate' and 'y-coordinate' double, so the point moves twice as far from the origin, but maintains the same 'bearing' with respect to the origin.
(d) $+1 \times i = i$, while $-1 \times i = -i$. In both cases, the multiplication by i seems to rotate the point by 90° anticlockwise about the origin. This is equally true if you start with $+2$ or -2 and multiply them by i.
(e) $iz = ix + i^2 y = ix - y$. The positions of z and iz are shown on the Argand diagram in Figure A.2, with the coordinates derived from our algebra. It follows from the geometry of Figure A.2 that angles $\angle$ZOR and $\angle$Z'OI are the same. Given that $\angle$IOR = 90°, it follows that $\angle$Z'OZ = 90°.

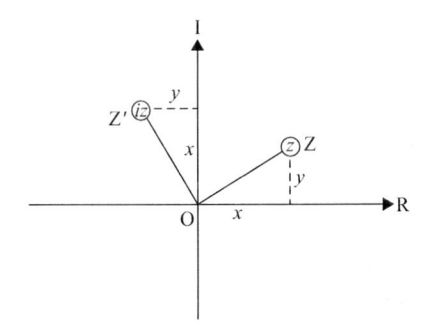

Fig. A.2

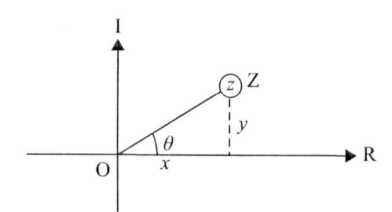

Fig. A.3

It is also apparent from the diagram that $OZ = OZ' = \sqrt{x^2 + y^2}$, so the transformation caused by multiplication by i is a pure rotation and does not involve any stretching or magnification.

(f) If $a = p + iq$ and $b = r + is$ then

$$ab = (p + iq)(r + is) = pr + iqr + ips + i^2 qs = pr - qs + i(qr + ps).$$

Thus $\operatorname{Re}(ab) = pr - qs = \operatorname{Re}(a)\operatorname{Re}(b) - \operatorname{Im}(a)\operatorname{Im}(b)$ and
$\operatorname{Im}(ab) = qr + ps = \operatorname{Im}(a)\operatorname{Re}(b) + \operatorname{Re}(a)\operatorname{Im}(b).$

(g) Reverse the sign of the imaginary part: $3 - 2i$, $5 + 3i$, $-2 - i$, $-4 + 5i$.

(h) If $a = p + iq$ and $b = r + is$, then $ab = pr - qs + i(qr + ps)$, while $a^*b^* = (p - iq)(r - is) = pr - qs - i(qr + ps)$.

We see that the real part of a^*b^* is the same as that of ab, while the imaginary part has reversed its sign. It follows that a^*b^* is the complex conjugate of ab.

(i) If $z = x + iy$, then $z* = x - iy$ and so $\frac{1}{2}(z + z^*) = \frac{1}{2}(2x) = x = \operatorname{Re}(z)$. Similarly $\frac{1}{2}(z - z^*) = \frac{1}{2}(2iy) = iy = i \times \operatorname{Im}(z)$. Thus $\operatorname{Im}(z) = \frac{1}{2i}(z - z^*)$.

(j) For $z = x + iy$, then we use Pythagoras' theorem to show that its 'direct distance' from the origin on the Argand diagram is $r = \sqrt{x^2 + y^2}$.
Evaluating z^*z we find $(x + iy)(x - iy) = x^2 + y^2 = r^2$ as requested.

(k) In Figure A.3, the modulus $|z|$ is represented by the length of the line OZ. $\operatorname{Re}(z) = x = |z|\cos\theta$, while $\operatorname{Im}(z) = y = |z|\sin\theta$, where $\theta = \operatorname{Arg}(z)$.

(l)
$$i^3 = i \times i^2 = i \times -1 = -i$$
$$i^4 = i \times i^3 = i \times -i = +1$$
$$i^5 = i \times i^4 = i \times 1 = +i$$
$$i^6 = i \times i^5 = i \times i = -1.$$

(m) Using our expression for e^x, it follows that

$$e^{i\theta} = 1 + i\theta + \frac{(i\theta)^2}{2!} + \frac{(i\theta)^3}{3!} + \frac{(i\theta)^4}{4!} + \frac{(i\theta)^5}{5!} + \frac{(i\theta)^6}{6!} + \cdots$$
$$= 1 + i\theta + \frac{i^2\theta^2}{2!} + \frac{i^3\theta^3}{3!} + \frac{i^4\theta^4}{4!} + \frac{i^5\theta^5}{5!} + \frac{i^6\theta^6}{6!} + \cdots$$

$$= 1 + i\theta - \frac{\theta^2}{2!} - \frac{i\theta^3}{3!} + \frac{\theta^4}{4!} + \frac{i\theta^5}{5!} - \frac{\theta^6}{6!} + \cdots$$

$$= \left(1 - \frac{\theta^2}{2!} + \frac{\theta^4}{4!} - \frac{\theta^6}{6!} + \cdots\right) + i\left(\theta - \frac{\theta^3}{3!} + \frac{\theta^5}{5!} + \cdots\right)$$

$$= \cos\theta + i\sin\theta.$$

(n) $re^{i\theta} \times se^{i\phi} = rs \times e^{i\theta}e^{i\phi} = rs\,e^{i(\theta+\phi)}$.
The modulus is rs, and the argument is $\theta + \phi$.

(o) $zw = |z||w|e^{i(\theta+\phi)}$, and since $|z| = |w| = 1$, $zw = e^{i(\theta+\phi)}$.
Thus $\mathrm{Arg}(zw) = \theta + \phi$, and $\mathrm{Re}(zw) = |zw|\cos(\mathrm{Arg}(zw)) = 1 \times \cos(\theta + \phi)$.
But from (f) we know that $\mathrm{Re}(zw) = \mathrm{Re}(z)\mathrm{Re}(w) - \mathrm{Im}(z)\mathrm{Im}(w)$.
Now $\mathrm{Re}(z) = \cos\theta$, $\mathrm{Im}(z) = \sin\theta$, $\mathrm{Re}(w) = \cos\phi$ and $\mathrm{Im}(w) = \sin\phi$.
Thus it follows that $\cos(\theta + \phi) = \cos\theta\cos\phi - \sin\theta\sin\phi$.
Similarly, $\mathrm{Im}(zw) = \sin(\mathrm{Arg}(zw)) = \sin(\theta + \phi)$ and since $\mathrm{Im}(zw) = \mathrm{Im}(z)\mathrm{Re}(w) + \mathrm{Re}(z)\mathrm{Im}(w)$ we have $\sin(\theta + \phi) = \sin\theta\cos\phi + \cos\theta\sin\phi$.

(p) From (m) we know that $e^{i\theta} = \cos\theta + i\sin\theta$. It follows that $e^{-i\theta} = \cos(-\theta) + i\sin(-\theta) = \cos\theta - i\sin\theta$.
Thus $\tfrac{1}{2}\left(e^{i\theta} + e^{-i\theta}\right) = \tfrac{1}{2}(2\cos\theta) = \cos\theta$, and $\tfrac{1}{2}\left(e^{i\theta} - e^{-i\theta}\right) = \tfrac{1}{2}(2i\sin\theta) = i\sin\theta$.

4.4 Damped oscillators

(a) The restoring force can be written $F = -kx$ (as in workshop 4.1.1), where the negative sign indicates that the force is directed in the opposite direction to x, and thus points back towards $x = 0$.
  The friction (or damping) force can be written $F = -r\dot{x}$, with the negative sign indicating that the force will always be in the opposite direction to the velocity (tending to decelerate it).
  Adding the two forces gives $F = -kx - r\dot{x}$, and so Newton's second law becomes:

$$F = \frac{d}{dt}m\dot{x} = m\ddot{x} = -kx - r\dot{x}, \quad \text{and thus } m\ddot{x} + kx + r\dot{x} = 0.$$

(b) If $x = \mathrm{Re}(Ae^{\alpha t})$ then $\dot{x} = \mathrm{Re}\left(A\alpha e^{\alpha t}\right)$ and $\ddot{x} = \mathrm{Re}\left(A\alpha^2 e^{\alpha t}\right)$. Our statement of Newton's law then becomes $m\,\mathrm{Re}\left(A\alpha^2 e^{\alpha t}\right) + r\,\mathrm{Re}\left(A\alpha e^{\alpha t}\right) + k\,\mathrm{Re}(Ae^{\alpha t}) = 0$.
This can be rewritten $\mathrm{Re}\left(Am\alpha^2 e^{\alpha t}\right) + \mathrm{Re}\left(Ar\alpha e^{\alpha t}\right) + \mathrm{Re}\left(Ake^{\alpha t}\right) = 0$ or

$$\mathrm{Re}\left(Am\alpha^2 e^{\alpha t} + Ar\alpha e^{\alpha t} + Ake^{\alpha t}\right) = 0$$
$$\mathrm{Re}\left\{\left(m\alpha^2 + r\alpha + k\right)Ae^{\alpha t}\right\} = 0$$
$$\mathrm{Re}\left\{\left(m\alpha^2 + r\alpha + k\right)z\right\} = 0.$$

This equation will definitely be satisfied if the stricter condition $(m\alpha^2 + r\alpha + k)z = 0$ is enforced.

(c) We solve the quadratic $m\alpha^2 + r\alpha + k = 0$ to give the solutions for α:

$$\alpha^2 + \frac{r}{m}\alpha = -\frac{k}{m}$$

$$\left(\alpha + \frac{r}{2m}\right)^2 = \left(\frac{r}{2m}\right)^2 - \frac{k}{m}$$

$$\alpha + \frac{r}{2m} = \pm\sqrt{\left(\frac{r}{2m}\right)^2 - \frac{k}{m}}$$

$$\alpha = -\frac{r}{2m} \pm \sqrt{\left(\frac{r}{2m}\right)^2 - \frac{k}{m}}$$

$$\alpha_1 = -\frac{r}{2m} + \sqrt{\left(\frac{r}{2m}\right)^2 - \frac{k}{m}}$$

$$\alpha_2 = -\frac{r}{2m} - \sqrt{\left(\frac{r}{2m}\right)^2 - \frac{k}{m}}.$$

(d) Our differential equation is $m\ddot{x} + r\dot{x} + kx = 0$. This will definitely be true if the more rigorous equation $m\ddot{z} + r\dot{z} + kz = 0$ is satisfied. If we use the solution suggested in equation (4.20) then:

$$z = e^{-rt/2m}(A_1 + A_2 t)$$

$$\dot{z} = e^{-rt/2m}\left(A_2 - \frac{r}{2m}A_1 - \frac{r}{2m}A_2 t\right)$$

$$\ddot{z} = e^{-rt/2m}\left(-\frac{r}{2m}A_2 - \frac{r}{2m}A_2 + \left(\frac{r}{2m}\right)^2 A_1 + \left(\frac{r}{2m}\right)^2 A_2 t\right).$$

Substituting these into the differential equation gives:

$$m\ddot{z} + r\dot{z} + kz = 0$$

$$e^{-rt/2m}\left(\left(-rA_2 + \frac{r^2}{4m}A_1 + \frac{r^2}{4m}A_2 t\right) + \left(rA_2 - \frac{r^2}{2m}A_1 - \frac{r^2}{2m}A_2 t\right)\right.$$

$$\left. + (kA_1 + kA_2 t)\right) = 0$$

$$e^{-rt/2m}\left(-\frac{r^2}{4m}A_1 - \frac{r^2}{4m}A_2 t + kA_1 + kA_2 t\right) = 0$$

but, since in our special case $r^2 = 4mk$, the term in brackets is equal to zero, and thus our solution is verified.

(e)
$$\alpha_1 = -\frac{r}{2m} + \sqrt{\left(\frac{r}{2m}\right)^2 - \frac{k}{m}} = -\frac{r}{2m} + \sqrt{(-1) \times \left(\frac{k}{m} - \left(\frac{r}{2m}\right)^2\right)}$$
$$= -\frac{r}{2m} + i\sqrt{\frac{k}{m} - \left(\frac{r}{2m}\right)^2}$$
$$\alpha_2 = -\frac{r}{2m} - \sqrt{\left(\frac{r}{2m}\right)^2 - \frac{k}{m}} = -\frac{r}{2m} - \sqrt{(-1) \times \left(\frac{k}{m} - \left(\frac{r}{2m}\right)^2\right)}$$
$$= -\frac{r}{2m} - i\sqrt{\frac{k}{m} - \left(\frac{r}{2m}\right)^2}.$$

(f) Let us define $\omega = \sqrt{(k/m) - (r/2m)^2}$. Since we are told $r^2 - 4mk < 0$, it follows that $4mk > r^2$ and the term in the square root is real, and so ω will be real as well.

Our solution z can then be written as

$$z = A_1 e^{\alpha_1 t} + A_2 e^{\alpha_2 t}$$
$$= A_1 e^{-rt/2m} e^{i\omega t} + A_2 e^{-rt/2m} e^{-i\omega t}$$
$$= e^{-rt/2m} \left(A_1 e^{i\omega t} + A_2 e^{-i\omega t} \right).$$

We can show that this can be expressed in terms of sines and cosines as follows, remembering from workshop 4.2 part (m) that $e^{i\theta} = \cos\theta + i\sin\theta$.

$$z = e^{-rt/2m} \left(A_1 e^{i\omega t} + A_2 e^{-i\omega t} \right)$$
$$= e^{-rt/2m} \left(A_1 \cos\omega t + iA_1 \sin\omega t + A_2 \cos(-\omega t) + A_2 \sin(-\omega t) \right)$$
$$= e^{-rt/2m} \left(A_1 \cos\omega t + iA_1 \sin\omega t + A_2 \cos\omega t - iA_2 \sin\omega t \right)$$
$$= e^{-rt/2m} \left((A_1 + A_2) \cos\omega t + i(A_1 - A_2) \sin\omega t \right)$$
$$x = \text{Re}(z) = e^{-rt/2m} \left(C \cos\omega t + D \sin\omega t \right),$$

where $C = \text{Re}(A_1 + A_2)$ and $D = -\text{Im}(A_1 - A_2)$.

(g)
$$x = e^{-at} \left(C \cos\omega t + D \sin\omega t \right)$$
$$\dot{x} = -ae^{-at} \left(C \cos\omega t + D \sin\omega t \right) + \omega e^{-at} \left(-C \sin\omega t + D \cos\omega t \right)$$
$$\ddot{x} = a^2 e^{-at} \left(C \cos\omega t + D \sin\omega t \right) - 2a\omega e^{-at} \left(-C \sin\omega t + D \cos\omega t \right)$$
$$+ \omega^2 e^{-at} \left(-C \cos\omega t - D \sin\omega t \right)$$
$$= \left(Ca^2 - 2Da\omega - C\omega^2 \right) e^{-at} \cos\omega t + \left(Da^2 + 2Ca\omega - D\omega^2 \right) e^{-at} \sin\omega t.$$

Inserting these trial solutions into the equation $m\ddot{x} + r\dot{x} + kx = 0$ gives us:

For the coefficients of $e^{-at} \cos\omega t$: $Cma^2 - 2Dma\omega - Cm\omega^2 - Car + Dr\omega + Ck = 0$.

For the coefficients of $e^{-at} \sin\omega t$: $Dma^2 + 2Cma\omega - Dm\omega^2 - Dar - Cr\omega + Dk = 0$.

Given that these equations need to be satisfied for any values of C or D, our cosine coefficient equation tells us that $ma^2 - m\omega^2 - ar + k = 0$ and $-2ma\omega + r\omega = 0$. From the second of these, we have $a = r/2m$, and substituting back into the first we have $(r^2/4m) - m\omega^2 - (r^2/2m) + k = 0$, so $-m\omega^2 - (r^2/4m) + k = 0$ and so $\omega = \sqrt{(k/m) - (r^2/4m^2)}$.

Analysis of the sine coefficient would have led to the same conclusion.

4.7 The wave equation (using trigonometry)

(a)
$$y = A\cos(\omega t - kx + \phi)$$

$$\frac{\partial y}{\partial x} = \frac{\partial y}{\partial(\omega t - kx + \phi)} \frac{\partial(\omega t - kx + \phi)}{\partial x}$$

$$= -A\sin(\omega t - kx + \phi) \times -k$$
$$= Ak\sin(\omega t - kx + \phi).$$

(b)
$$\frac{\partial^2 y}{\partial x^2} = \frac{\partial(\partial y/\partial x)}{\partial(\omega t - kx + \phi)} \frac{\partial(\omega t - kx + \phi)}{\partial x}$$

$$= Ak\cos(\omega t - kx + \phi) \times -k$$
$$= -Ak^2\cos(\omega t - kx + \phi) = -k^2 y.$$

(c)
$$y = A\cos(\omega t - kx + \phi)$$

$$\frac{\partial y}{\partial t} = \frac{\partial y}{\partial(\omega t - kx + \phi)} \frac{\partial(\omega t - kx + \phi)}{\partial t}$$

$$= -A\sin(\omega t - kx + \phi) \times \omega$$
$$= -A\omega\sin(\omega t - kx + \phi).$$

(d)
$$\frac{\partial^2 y}{\partial t^2} = \frac{\partial(\partial y/\partial t)}{\partial(\omega t - kx + \phi)} \frac{\partial(\omega t - kx + \phi)}{\partial t}$$

$$= -A\omega\cos(\omega t - kx + \phi) \times \omega$$
$$= -A\omega^2\cos(\omega t - kx + \phi) = -\omega^2 y.$$

(e) $(\partial^2 y/\partial x^2) = -k^2 y$, and $(\partial^2 y/\partial t^2) = -\omega^2 y$.
It follows that $-y = (1/k^2)(\partial^2 y/\partial x^2) = (1/\omega^2)(\partial^2 y/\partial t^2)$
and thus that $(\partial^2 y/\partial x^2) = (k^2/\omega^2)(\partial^2 y/\partial t^2)$.
Now $\omega/k = c$, so $(\partial^2 y/\partial x^2) = (1/c^2)(\partial^2 y/\partial t^2)$.

(f)
$$y = f(\omega t - kx)$$

$$\frac{\partial y}{\partial x} = \frac{\partial y}{\partial(\omega t - kx)} \frac{\partial(\omega t - kx)}{\partial x}$$
$$= f' \times -k$$

$$\frac{\partial y}{\partial t} = \frac{\partial y}{\partial (\omega t - kx)} \frac{\partial (\omega t - kx)}{\partial t}$$
$$= f' \times \omega$$
$$\frac{\partial^2 y}{\partial x^2} = \frac{\partial (\partial y/\partial x)}{\partial (\omega t - kx)} \frac{\partial (\omega t - kx)}{\partial x}$$
$$= -kf'' \times -k = k^2 f''$$
$$\frac{\partial^2 y}{\partial t^2} = \frac{\partial (\partial y/\partial t)}{\partial (\omega t - kx)} \frac{\partial (\omega t - kx)}{\partial t}$$
$$= \omega f'' \times \omega = \omega^2 f''.$$

where f' refers to the derivative of the function f with respect to its argument. Similar working using $\omega t + kx$ gives identical results for the second derivatives.

It is clear from the derivatives above that $(1/k^2)(\partial^2 y/\partial x^2) = (1/\omega^2)(\partial^2 y/\partial t^2)$, and so the wave equation is satisfied.

4.7 The wave equation (using complex numbers)

(a)
$$y = \text{Re } Ae^{i\phi}e^{i(\omega t - kx)}$$
$$\frac{\partial y}{\partial x} = \frac{\partial y}{\partial (\omega t - kx)} \frac{\partial (\omega t - kx)}{\partial x}$$
$$= Ae^{i\phi}e^{i(\omega t - kx)} \times -ik$$
$$= -Aike^{i\phi}e^{i(\omega t - kx)}.$$

(b)
$$\frac{\partial^2 y}{\partial x^2} = \frac{\partial (\partial y/\partial x)}{\partial (\omega t - kx)} \frac{\partial (\omega t - kx)}{\partial x}$$
$$= -Aike^{i\phi}e^{i(\omega t - kx)} \times -ik$$
$$= -Ak^2 e^{i\phi}e^{i(\omega t - kx)} = -k^2 y.$$

(c)
$$y = \text{Re } Ae^{i\phi}e^{i(\omega t - kx)}$$
$$\frac{\partial y}{\partial t} = \frac{\partial y}{\partial (\omega t - kx)} \frac{\partial (\omega t - kx)}{\partial t}$$
$$= Ae^{i\phi}e^{i(\omega t - kx)} \times i\omega$$
$$= Ai\omega e^{i\phi}e^{i(\omega t - kx)}.$$

(d)
$$\frac{\partial^2 y}{\partial t^2} = \frac{\partial (\partial y/\partial t)}{\partial (\omega t - kx)} \frac{\partial (\omega t - kx)}{\partial t}$$
$$= Ai\omega e^{i\phi}e^{i(\omega t - kx)} \times i\omega$$
$$= -A\omega^2 e^{i\phi}e^{i(\omega t - kx)} = -\omega^2 y.$$

(e) & (f) See solution presented above after earlier parts have been solved using trigonometry.

4.11.3 Stellar magnitudes

(a) Let us suppose the intensity of a sixth-magnitude star ($m = 6.0$) is called I_6, and we define a constant a such that $I_5 = aI_6$. It follows by the constant-ratio rule that $I_4 = aI_5 = a^2 I_6$, and so on, until we have that $I_1 = a^5 I_6$. We are told in the text that an increase of five in the magnitude gives a 100-fold reduction in I, thus $I_1 = 100\, I_6$, and so $a^5 = 100$. It follows that $a = \sqrt[5]{100} = 100^{1/5} = 10^{2/5} = 2.512\ldots$, and given that $I_1 = a\, I_2$, a first magnitude star is 2.512 times as bright as a second magnitude star.

(b) Suppose the brightness of a zero-magnitude star ($m = 0$) is called I_0. It follows from the logic of (a) that $I_m = I_0/a^m = I_0 a^{-m} = I_0 \left(10^{2/5}\right)^{-m} = I_0 \times 10^{-2m/5}$.

(c) From part (b) we have that $I \propto 10^{-2m/5}$ while $P \propto 10^{-2M/5}$. Given that the 4π factor is a constant, we can write $10^{-2m/5} \propto I \propto P/D^2 \propto 10^{-2M/5}/D^2$. Defining B as the constant of proportionality, we have $10^{-2m/5} = B \times 10^{-2M/5}/D^2$.

(d) When $D = 10\,\text{pc}$, $M = m$. Putting $M = m$ into our equation from part (c) gives us $1 = B/D^2 = B/100\,\text{pc}^2$, and so $B = 100\,\text{pc}^2$.

(e) Using logarithms, as requested

We take logarithms to base 10 of both sides, and assume that D is in parsecs. This means that $B = 100\,\text{pc}^2$, and hence $\log B = 2$.

$$10^{-2m/5} = \frac{B \times 10^{-2M/5}}{D^2}$$

$$-\frac{2m}{5} = \log B - \frac{2M}{5} - 2\log D$$

$$-\frac{2m}{5} = 2 - \frac{2M}{5} - 2\log D$$

$$-m = 5 - M - 5\log D$$

$$\log D = \frac{5 + m - M}{5}$$

$$D = 10^{(5+m-M)/5}.$$

(e) Without using logarithms
While doing this using logarithms is good practice, it is actually easier to do it without – and here is how:

$$10^{-2m/5} = \frac{B \times 10^{-2M/5}}{D^2}$$

$$D^2 = \frac{B \times 10^{-2M/5}}{10^{-2m/5}} = B \times 10^{-2(M-m)/5}$$

$$= 100 \times 10^{-2(M-m)/5} = 10^{2-2(M-m)/5}$$

$$D = 10^{1-(M-m)/5} = 10^{(5+m-M)/5}$$

Chapter 5

5.1.3 Using voltage to solve simple circuit problems

Voltage labelling (see key at base for symbols used) (Figure A.4)
 Potential differences (all in volts) and bulb brightnesses (bright/normal/dim/off)

No.	Potential difference		No.	Potential difference		No.	Potential difference	
1	$1.5 - 0 = 1.5$	N	9	$1.5 - 0 = 1.5$	N	17	$3 - 1.5 = 1.5$	N
2	$3 - 0 = 3$	B	10	$1.5 - 0 = 1.5$	N	18	$3 - 3 = 0$	O
3	$1.5 - 0 = 1.5$	N	11	$0.75 - 0 = 0.75$	D	19	$1.5 - 0 = 1.5$	N
4	$1.5 - 0 = 1.5$	N	12	$1.5 - 0.75 = 0.75$	D	20	$3 - 1.5 = 1.5$	N
5	$0.75 - 0 = 0.75$	D	13	$0.75 - 0 = 0.75$	D	21	$1.5 - 1.5 = 0$	O
6	$1.5 - 0.75 = 0.75$	D	14	$1.5 - 0.75 = 0.75$	D	22	$1.5 - 0 = 1.5$	N
7	$2.25 - 0 = 2.25$	B	15	$1.5 - 0 = 1.5$	N	23	$3 - 1.5 = 1.5$	N
8	$4.5 - 2.25 = 2.25$	B	16	$1.5 - 0 = 1.5$	N			

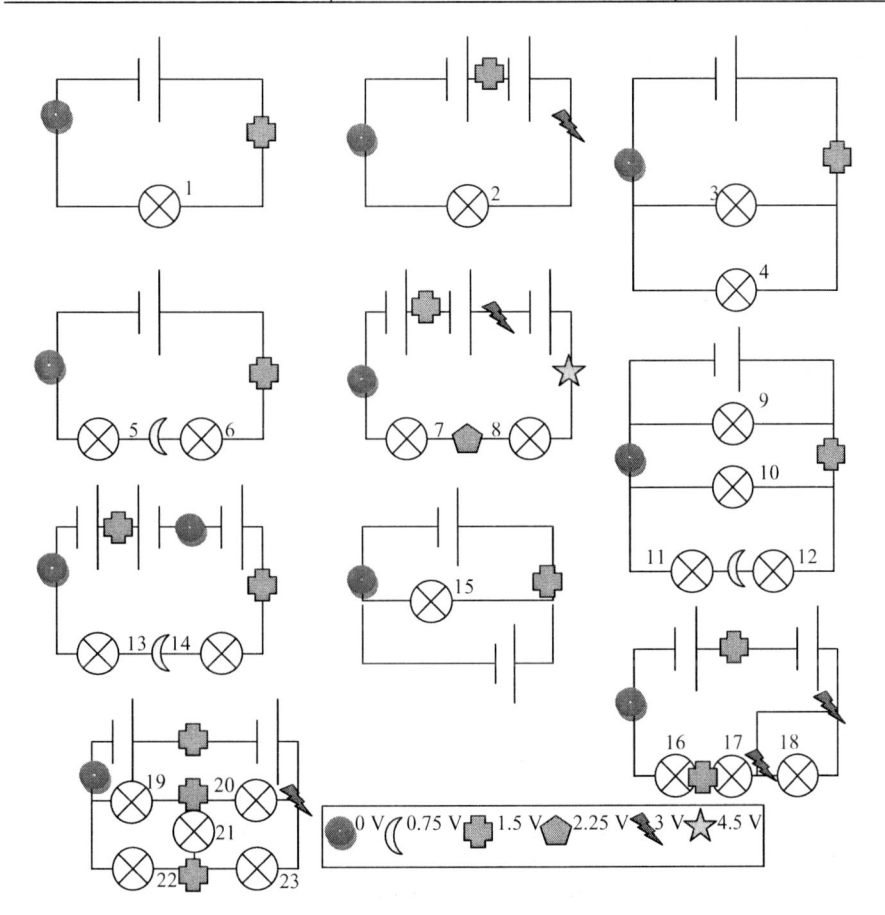

Fig. A.4

Chapter 6

6.5.1 Atmospheric pressure

(a)
$$\rho = \frac{\text{Mass}}{\text{Volume}} = \frac{\text{Number of molecules} \times \text{mass of molecule}}{\text{Volume}}$$
$$= \frac{Nm}{V} = \frac{pV}{kT}\frac{m}{V} = \frac{pm}{kT}.$$

(b) Mass of one slab = volume × density = $V\rho = \Delta h \rho$
(since slabs have unit area).

Weight of one slab = mass × $g = \rho g \Delta h = \frac{pm}{kT} g \Delta h = \frac{pmg}{kT} \Delta h.$

Difference in force supporting this slab compared to force needed to support next slab up will be equal to this weight. Now because the slabs have unit area, the difference in the force is numerically equal to the difference in pressure. Thus the difference in pressure as we go up a height Δh is $(pmg/kT)\Delta h$, where the pressure reduces as we rise. Thus we write $\Delta p = -(pmg/kT)\Delta h$.

(c) If we now let the height increments become small, then $(\Delta p/\Delta h) = -(pmg/kT)$ tends to $(dp/dh) = -(pmg/kT) = -(mg/kT)p$. This is a differential equation of the form $(dy/dx) = ky$, and workshop 7.2 shows us that the solution is $y = Ae^{kx}$, where A is a constant. Applying this to our equation, we have $p = p_0 e^{-mgh/kT}$, where p_0 is the pressure at height $h = 0$. For the final part, we note that $mgh = E_{\text{grav}}$ is the potential energy gained by a molecule when it rises by height h, and we can therefore write $p = p_0 e^{-E_{\text{grav}}/kT}$.

6.5.3 Justification of Boltzmann law

(a)

Macrostate	Number of microstates	Microstates
3 Heads	1	HHH
2 Heads, 1 Tail	3	THH, HTH, HHT
1 Head, 2 Tails	3	HTT, THT, TTH
3 Tails	1	TTT

(b) & (c)
- Two atoms with two units, five atoms with no units {5,0,2}
- One atom with two units, two with one unit, four with no units {4,2,1}
- One atom with four units, six with none {6,0,0,0,1}.

(d) Using first principles in each case

{6,0,0,0,1} Seven microstates (we choose one of the seven atoms).

{5,1,0,1} We choose one of the seven to hold three units, and then choose one of the remaining six to hold the remaining unit. This means $7 \times 6 = 42$ microstates.

{5,0,2} We choose one atom from 7, then choose the second from the remaining 6. This gives us 7×6 possible ways of doing this, but in doing so we double count each possibility (choosing atom 3 then atom 5 is the same as choosing atom 5 then atom 3). Thus the number of different microstates is $7 \times 6/2 = 21$.

{4,2,1} We choose one atom from the seven to hold two units: there are seven to choose from. We then choose two from the remaining six to take the other individual units: there are $6 \times 5/2 = 15$ ways of doing this. Thus the number of microstates is $7 \times 15 = 105$.

{3,4} Here we need to choose three of the atoms to take no energy. We start by choosing the first from 7, then choose the second from the remaining 6, then the third from 5. This means $7 \times 6 \times 5 = 210$ possibilities, but we have counted each possible microstate six times over. To see why, suppose the microstate is to choose atoms 1, 2 and 3. We could choose (1 then 2 then 3) or (1 then 3 then 2) or (2 then 3 then 1) or (2 then 1 then 3) or (3 then 1 then 2) or (3 then 2 then 1). Thus the number of different microstates is $210/6 = 35$.

(d) Using a general procedure

In this solution, we shall justify the equation given in footnote 7 of Chapter 6 and then use it.

Proof

(i) Suppose you have N objects. The number of ways of putting them in order equals $N!$, that is $N \times (N-1) \times (N-2) \ldots \times 2 \times 1$. To see why, you first have to choose one of the N to go first (there are N to choose from). You then have to choose one of the $(N-1)$ remaining to go second (there are $(N-1)$ to choose from), one of the $(N-2)$ to go third, and so on, until you have no choice when it come to the last object.

(ii) We want to work out how many ways there are of assigning the energy units to the atoms for a particular macrostate. To do this, we shall prepare a grid to hold the atoms. The first n_0 places in the grid are reserved for the atoms which will end up with no energy, the next n_1 places will hold the atoms which are to have 1 unit of energy, and so on. There are $N!$ ways of putting the atoms in the ordered grid (as discussed above). However this does not mean that there are $N!$ microstates, since any rearrangement of the n_0 atoms, say, within their places on the grid does not change the microstate since the same atoms each time will have no energy, they will just go in a different order within the n_0 places at the beginning of the grid. There are, of course, $n_0!$ ways of rearranging these atoms within their places on the grid. We therefore have to divide our $N!$ ways by $n_0!$ to avoid counting these rearrangements of the no-energy atoms as different microstates. Similarly we also have to divide by $n_1!$ to avoid counting rearrangements of the atoms with 1 unit as different microstates. We also have to divide by $n_2!$, $n_3!$, and so on. This gives us the equation for the number of microstates in the macrostate $\{n_0, n_1, n_2, n_3, \ldots\}$ as $\dfrac{N!}{n_0! n_1! n_2! \ldots}$.

Evaluation

Macrostate	Calculation
$\{6,0,0,0,1\}$	$\dfrac{7!}{6!1!} = \dfrac{7 \times 6 \times 5 \times 4 \times 3 \times 2 \times 1}{(6 \times 5 \times 4 \times 3 \times 2 \times 1) \times 1} = 7$
$\{5,1,0,1\}$	$\dfrac{7!}{5!1!1!} = \dfrac{7 \times 6 \times 5 \times 4 \times 3 \times 2 \times 1}{(5 \times 4 \times 3 \times 2 \times 1) \times 1 \times 1} = 7 \times 6 = 42$
$\{5,0,2\}$	$\dfrac{7!}{5!2!} = \dfrac{7 \times 6 \times 5 \times 4 \times 3 \times 2 \times 1}{(5 \times 4 \times 3 \times 2 \times 1) \times (2 \times 1)} = \dfrac{7 \times 6}{2} = 21$
$\{4,2,1\}$	$\dfrac{7!}{4!2!1!} = \dfrac{7 \times 6 \times 5 \times 4 \times 3 \times 2 \times 1}{(4 \times 3 \times 2 \times 1) \times (2 \times 1) \times 1} = \dfrac{7 \times 6 \times 5}{2} = 105$
$\{4,3\}$	$\dfrac{7!}{4!3!} = \dfrac{7 \times 6 \times 5 \times 4 \times 3 \times 2 \times 1}{(4 \times 3 \times 2 \times 1) \times (3 \times 2 \times 1)} = \dfrac{7 \times 6 \times 5}{6} = 35$

(e) There is a clear winner – $\{4,2,1\}$ and the numbers in the curly brackets happen to form a geometric series, each one being half of the one before.

But why is the most likely macrostate the one with the geometric progression? A hint as to the method of proof was given in the footnote 7 of Chaper 6, we shall expand this here.

We start by assuming that we have a geometric progression, and so we may write $n_p = n_0 f^p$. Accordingly, the number of microstates in the macrostates is given by

$$W = \frac{N!}{n_0! n_1! n_2! \ldots} = \frac{N!}{n_0! (n_0 f)! (n_0 f^2)! \ldots}.$$

We now move one of the units of energy from an atom that had p units to one that has q. The atom that had p now has $p - 1$, and the atom that had q now has $q + 1$. Thus n_p and n_q have gone down by one, while n_{p-1} and n_{q+1} have gone up by one. To avoid confusion, we shall use n_p (and so on) to refer to the number of atoms with p units of energy *before* the swap, and use primed letters (n'_p) to refer to the situation *after* the swap. This means that

n_p has gone down by 1 so $n'_p! = (n_p - 1)! = \dfrac{n_p!}{n_p}$

n_q has gone down by 1 so $n'_q! = (n_q - 1)! = \dfrac{n_q!}{n_q}$

n_{p-1} has gone up by 1 so $n'_{p-1}! = (n_{p-1} + 1)! = n_{p-1}! \times (n_{p-1} + 1)$
n_{q+1} has gone up by 1 so $n'_{q+1}! = (n_{q+1} + 1)! = n_{q+1}! \times (n_{q+1} + 1)$.

This in turn means that the number of microstates in the new macrostate is given by

$$W' = \frac{N!}{n_0'!n_1'!n_2'!\cdots} = W\frac{n_p n_q}{(n_{p-1}+1)(n_{q+1}+1)}.$$

Now $n_{q+1}/n_q = n_p/n_{p-1} = f$, and so we have:

$$W' = W\frac{n_p}{n_{p-1}+1}\frac{n_q}{n_{q+1}+1} = W \times \frac{f}{1+\frac{1}{n_{p-1}}} \times \frac{1}{f+\frac{1}{n_q}}$$

$$= W \times \frac{1}{1+\frac{1}{n_{p-1}}} \times \frac{1}{1+\frac{1}{fn_q}} < W.$$

Thus the new state has fewer microstates than the original one. This indicates that the original macrostate was the one with the most microstates.* In short, the geometric series distribution has the most microstates, and if the microstates are equally likely (a workable and useful assumption), this will be the one most likely to occur.

(f) $e^{-\alpha p} = (e^{-\alpha})^p$, so $f = e^{-\alpha}$.

(g) Given that $\sum\limits_p n_p = N$, it follows that

$$N = \sum_{p=0}^{\infty} n_p = \sum_{p=0}^{\infty} Af^p = A\sum_{p=0}^{\infty} f^p.$$ As suggested in the hint, we notice that

$$Nf = f\sum_{p=0}^{\infty} n_p = \sum_{p=0}^{\infty} Af^{p+1} = A\sum_{q=1}^{\infty} f^q = A\sum_{q=0}^{\infty} f^q - Af^0 = N - A.$$

Thus:

$$Nf = N - A$$
$$A = N(1-f)$$
$$N = \frac{A}{1-f}.$$

Please note in passing that this means that $\sum\limits_{p=0}^{\infty} f^p = 1+f+f^2+f^3+\cdots = \frac{1}{1-f}$, a fact which we shall be using in our next part.

(h) The total number of energy units:

$$E = \sum_{p=0}^{\infty} pn_p = A\sum_{p=0}^{\infty} pf^p = A\left(f + 2f^2 + 3f^3 + 4f^4 + \cdots\right).$$

---

*As it stands, this does not constitute a formal proof that the 'geometric progression distribution' has a global maximum on the number of microstates. In addition, the cases where $p = q$ or $p = q+1$ require special consideration as the swaps then 'cross'. However the tidying up of the situation does not involve any new physics, and as far as we are concerned, this ought to give you as good a reason as any why the geometric series distribution is the one favoured by nature.

202   Workshop solutions

As suggested in the hint, the quantity in brackets can be expressed:

$$f + 2f^2 + 3f^3 + 4f^4 + \cdots = \begin{array}{l} f\ +f^2\ +f^3\ +f^4+\cdots \\ +f^2\ +f^3\ +f^4+\cdots \\ +f^3\ +f^4+\cdots \\ +f^4+\cdots \end{array}$$

$$= \left(f + f^2 + f^3 + f^4 + \cdots\right)$$
$$+ f\left(f + f^2 + f^3 + \cdots\right)$$
$$+ f^2\left(f + f^2 + \cdots\right)$$
$$+ f^3\left(f + \cdots\right)$$
$$= \left(1 + f + f^2 + f^3 + f^4 + \cdots\right)$$
$$\times \left(f + f^2 + f^3 + f^4 + \cdots\right)$$
$$= \frac{1}{1-f} \times f\left(\frac{1}{1-f}\right) = \frac{f}{(1-f)^2}.$$

and so $E = \dfrac{Af}{(1-f)^2}$.

(i) From part (g) we know $A = N(1-f)$. Inserting this into our answer to part (h) gives:

$$E = \frac{Af}{(1-f)^2} = \frac{N(1-f)f}{(1-f)^2} = \frac{Nf}{(1-f)}, \quad \text{so } E(1-f) = Nf$$

and hence:

$$E = f(E+N) \quad \text{and} \quad f = \frac{E}{E+N} = \frac{\varepsilon}{\varepsilon+1}$$

using $\varepsilon = E/N$ as suggested.

We can now write $A = N(1-f) = N\left(1 - \dfrac{\varepsilon}{\varepsilon+1}\right) = N\dfrac{1}{\varepsilon+1}$, and we have equations for A and f in terms of N and E. We then substitute to find:

$$n_p = Af^p = \frac{N}{\varepsilon+1}\left(\frac{\varepsilon}{\varepsilon+1}\right)^p = \frac{N}{\varepsilon+1}\left(\frac{1}{1+\frac{1}{\varepsilon}}\right)^p = \frac{N}{\varepsilon+1}\left(1+\varepsilon^{-1}\right)^{-p}.$$

(j) As shown in workshop 7.2, we may write e^x as an expansion $1 + x + \dfrac{x^2}{2!} + \dfrac{x^3}{3!} + \cdots$, and hence $f^{-1} = e^{\alpha} = 1 + \alpha + \dfrac{\alpha^2}{2!} + \dfrac{\alpha^3}{3!} + \cdots$. From part (i) we know that $f^{-1} = 1 + \varepsilon^{-1}$. Thus $1 + \varepsilon^{-1} = 1 + \alpha + \dfrac{\alpha^2}{2!} + \dfrac{\alpha^3}{3!} + \cdots$. If ε is very large, then ε^{-1} will be very small, and therefore $1 + \varepsilon^{-1}$ will be very close to 1. In this case, we can get a good approximation to the series by truncating it: $1 + \varepsilon^{-1} = 1 + \alpha$, and so we approximate $\alpha = \varepsilon^{-1}$. With this done, $n_p = Ae^{-\alpha p} = Ae^{-p/\varepsilon}$.

(k) If an atom has F joules of energy, then $p = F/\eta$. We can also write $\varepsilon = \bar{F}/\eta$. Thus $n_p = A\exp\left(-(F/\eta)/(\bar{F}/\eta)\right) = Ae^{-F/\bar{F}}$. So if $\bar{F} = kT$, then it follows that the probability that an atom will have energy F is given by $n_p = Ae^{-F/kT}$, which is, of course, the Boltzmann law.

Note: You may be surprised by the complexity of the argument in this workshop and by the number of assumptions we have had to make. However these assumptions are made in modern statistical mechanics, and prove to be excellent descriptions of nature. Beware simpler treatments – the assumptions may be still present, but hidden.

Chapter 7
7.1 Setting up integrals

(a) From more elementary work we remember that the moment of a force about a pivot is the force times the perpendicular distance from the pivot. Choosing our line AB to be the pivot we have that the moment of a force applied by the water to the dam at a depth y is

$$\Delta\Gamma = \text{Pressure} \times \text{area} \times (h-y) = \rho g y \times l\Delta y \times (h-y).$$

Now, once again

$$\Delta\Gamma = \frac{d\Gamma}{dy}\Delta y$$

is an approximate expression for $\Delta\Gamma$ for small Δy and the approximation gets better as $\Delta y \to 0$, so in this limit:

$$\frac{d\Gamma}{dy} = \rho g l y(h-y).$$

So, the total resultant moment Γ is the limiting sum:

$$\Gamma = \int_0^h \frac{d\Gamma}{dy}dy = \rho g l \int_0^h y(h-y)dy = \frac{\rho g l h^3}{6}.$$

(b) From the preamble of the workshop the resultant force F was

$$F = \int_0^h \frac{dF}{dy}dy = \rho g l \int_0^h y\,dy = \frac{\rho g l h^2}{2},$$

so

$$F \times (h - y_o) = \frac{\rho g l h^3}{6},$$

and therefore $h - y_o = h/3$; that is, F would have to act at a height $h/3$ from AB in order to apply the same moment as the combined turning effect of all the water on the dam.

(c) The volume integral is broken down into a product of three other integrals:

$$V = \int_0^R r^2 dr \int_0^\pi \sin\theta d\theta \int_0^{2\pi} d\varphi.$$

Performing each one separately,

$$V = \frac{R^3}{3} \times 2 \times 2\pi = \frac{4}{3}\pi R^3.$$

7.2 Logarithms

(a) $\log 100\,000 = \log 10^5 = 5$
$\log 10\,000 = \log 10^4 = 4$
$\log(10^5 \times 10^4) = \log 10^9 = 9 = 5 + 4$.
Thus $\log(ab) = \log a + \log b$.

(b) $\log(10^5 \div 10^4) = \log 10 = \log 10^1 = 1 = 5 - 4$.
Thus $\log(a/b) = \log a - \log b$.

(c) (i) $6 = 2 \times 3$, so $\log 6 = \log 2 + \log 3 = 0.778$
(ii) $1.5 = 3 \div 2$ so $\log 1.5 = \log 3 - \log 2 = 0.176$
(iii) $4 = 2 \times 2$ so $\log 4 = \log 2 + \log 2 = 2 \times \log 2 = 0.602$
(iv) $9 = 3 \times 3 = 3^2$ so $\log 9 = 2 \times \log 3 = 0.954$
(v) $0.5 = 1 \div 2$ so $\log 0.5 = \log 1 - \log 2 = 0 - \log 2 = -0.301$
or $0.5 = 2^{-1}$ so $\log 0.5 = -1 \times \log 2 = -0.301$.

(d) Taking logs of both sides of $y = 100x^4$, we get:

$$\log y = \log 100 + \log x^4$$
$$= \log 100 + \log(xxxx)$$
$$= \log 100 + \log x + \log x + \log x + \log x$$
$$= \log 100 + 4\log x$$
$$\log y = 2 + 4\log x.$$

Thus when we plot $\log y$ against $\log x$ we get a straight line with gradient 4 and y-intercept 2.

(e) Each time another half-life passes, m needs to halve so it must be multiplied by another factor of $\frac{1}{2}$. The number of half-lives passed is t/T, so we need to multiply m_0 (the starting mass) by one half t/T times. Thus $m = m_0 \left(\frac{1}{2}\right)^{t/T}$, and accordingly:

$$\log m = \log m_0 + \log\left(\frac{1}{2}\right)^{t/T}$$
$$= \log m_0 + \frac{t}{T}\log\frac{1}{2}$$
$$= \log m_0 - \frac{t}{T}\log 2,$$

so we get a straight line when we plot $\log m$ against t, with gradient $-\log 2/T$ and y-intercept $\log m_0$.

(f)
$$56 = 2^x$$
$$\log 56 = \log 2^x = x \log 2$$
$$x = \frac{\log 56}{\log 2} = \frac{1.748}{0.310} = 5.81.$$

(g)
$$\ln ab = \int_1^{ab} \frac{1}{y} dy = \int_1^a \frac{1}{y} dy + \int_a^{ab} \frac{1}{y} dy = \ln a + \int_a^{ab} \frac{1}{y} dy.$$

We now use the substitution $y = az$ to evaluate the remaining integral:

$$\ln a + \int_a^{ab} \frac{1}{y} dy = \ln a + \int_{z=1}^b \frac{1}{az} d(az) = \ln a + \int_{z=1}^b \frac{1}{az} a\, dz = \ln a + \int_{z=1}^b \frac{1}{z} dz$$
$$= \ln a + \ln b.$$

(h)
$$\frac{d}{dx} Ae^x = A \frac{d}{dx} e^x = Ae^x \quad \text{as required.}$$

(i)
$$\frac{d}{dx} Ae^{kx} = A \frac{d}{dx} e^{kx} = A \frac{de^{kx}}{d(kx)} \frac{d(kx)}{dx} = Ae^{kx} k = kAe^{kx}.$$

Thus if $y = Ae^{kx}$, then $\frac{dy}{dx} = kAe^{kx} = ky$, so $y = Ae^{kx}$ is the solution to the differential equation $\frac{dy}{dx} = ky$.

(j) (i) If $x = 0$, then $e^x = e^0 = a_0$. Now anything raised to the 0th power is one, so thus $a_0 = 1$.

(ii) If $y = a_0 + a_1 x + a_2 x^2 + a_3 x^3 + \cdots + a_p x^p + \cdots$ then it follows that $\frac{dy}{dx} = a_1 + 2a_2 x + 3a_3 x^2 + \cdots + pa_p x^{p-1} + \cdots$. Now if we require that $y = \frac{dy}{dx}$, then the coefficients for each power of x must be the same in the two equations. So it must be true that $a_0 = a_1$, $a_1 = 2a_2$, $a_2 = 3a_3$, and so on.

Thus $a_{p-1} = pa_p$.

(iii) From (ii) we have that $a_p = (a_{p-1}/p)$. We know that $a_0 = 1$, so $a_1 = a_0/1 = 1$. Next, $a_2 = \frac{a_1}{2} = \frac{1}{2}$, $a_3 = \frac{a_2}{3} = \frac{1/2}{3} = \frac{1}{2 \times 3} = \frac{1}{3!}$, $a_4 = \frac{a_3}{4} = \frac{1/3!}{4} = \frac{1}{3! \times 4} = \frac{1}{4!}$, and it follows that $a_p = \frac{a_{p-1}}{p} = \frac{1/(p-1)!}{p} = \frac{1}{(p-1)! \times p} = \frac{1}{p!}$.

$$e^x = a_0 + a_1 x + a_2 x^2 + a_3 x^3 + \cdots + a_p x^p + \cdots$$

Thus
$$= 1 + x + \frac{x^2}{2!} + \frac{x^3}{3!} + \cdots + \frac{x^p}{p!} + \cdots.$$

(iv) We now substitute $x = 1$ into our equation and get:

$$e = e^1 = 1 + 1 + \frac{1}{2!} + \frac{1}{3!} + \cdots + \frac{1}{p!} + \cdots$$
$$= 2.0000 + 0.5000 + 0.1667 + 0.0417 + 0.0083 + 0.0014 + 0.0002$$
$$= 2.718$$

to four significant figures.

7.3 Rockets and stages

(a) The conservation of momentum would tell us that

$$(m + M)v = m(v - v_e) + M(v + \Delta v);$$

that is, the momentum of the rocket + element of fuel before the fuel is ejected must equal the momentum of the rocket + the momentum of the element of fuel after the fuel is ejected. The ejection of fuel leads to a small increment in velocity Δv along the direction of motion of the rocket. Expanding this expression and cancelling common terms on both sides of the equality sign leads to

$$0 = -mv_e + M\Delta v.$$

Using the suggested notation in the workshop, $m = -\Delta M$ we immediately get:

$$\Delta v = -\frac{v_e}{M}\Delta M,$$

which gives an approximation for the velocity increment Δv that gets better as $\Delta M \to 0$. So in the limit, the sum of all the velocity increments from a velocity v_0 up to a velocity v_1 becomes:

$$\int_{v_0}^{v_1} dv = \int_{M_0}^{M_1} \frac{dv}{dM} dM = -\int_{M_0}^{M_1} \frac{v_e}{M} dM.$$

(b) Performing the integration:

$$[v]_{v_0}^{v_1} = -v_e \left[\ln(M)\right]_{M_0}^{M_1},$$

and introducing the limits:

$$v_1 - v_0 = -v_e(\ln(M_1) - \ln(M_0)) = v_e \ln\left(\frac{M_0}{M_1}\right).$$

(c) From the descriptions of s and l we get:

$$s = \frac{M_S}{M_0 - M_L} \quad \text{and} \quad l = \frac{M_L}{M_0}.$$

Now:

$$\frac{M_1}{M_0} = \frac{M_S + M_L}{M_0} = \frac{M_S}{M_0} + l = \frac{M_S(M_0 - M_L)}{(M_0 - M_L)M_0} + l = s(1-l) + l,$$

so

$$v_1 - v_0 = -v_e \ln\left(s(1-l) + l\right).$$

(d) If all the parameters v_e, s, and l are the same for every stage, then the velocity increment is the same for every stage, therefore:

$$v_1 - v_0 = -v_e N \ln\left(s(1-l) + l\right)$$

for N stages. However, this expression can be misleading. Let us analyse more carefully what happens as $N \to \infty$:

(e) $\lambda = M_L/M_0^{(1)}$
$$= (M_0^{(2)}/M_0^{(1)}) \times (M_0^{(3)}/M_0^{(2)}) \times (M_0^{(4)}/M_0^{(3)}) \times \cdots \times (M_L/M_0^{(N-1)}),$$

so $\lambda = l^N$, which means that $l = \lambda^{1/N}$, we have:

$$v_1 - v_0 = -N v_e \ln\left(s(1-l) + l\right) = -N v_e ln(s(1 - \lambda^{1/N}) + \lambda^{1/N}).$$

(f)
$$N \ln\left(s(1 - \lambda^{\frac{1}{N}}) + \lambda^{\frac{1}{N}}\right) = N \ln\left(\lambda^{\frac{1}{N}}\left(1 + \frac{s(1 - \lambda^{\frac{1}{N}})}{\lambda^{\frac{1}{N}}}\right)\right)$$

$$= \ln(\lambda) + N \ln\left(1 + \frac{s(1 - \lambda^{\frac{1}{N}})}{\lambda^{\frac{1}{N}}}\right).$$

Using the first expansion provided:

$$1 - \lambda^{\frac{1}{N}} = -\frac{1}{N} \ln \lambda - \frac{1}{2!}\left(\frac{1}{N} \ln \lambda\right)^2 - \frac{1}{3!}\left(\frac{1}{N} \ln \lambda\right)^3 + \cdots,$$

so

$$\frac{s(1 - \lambda^{\frac{1}{N}})}{\lambda^{\frac{1}{N}}} = -\frac{s}{N\lambda^{\frac{1}{N}}} \ln \lambda - \frac{s}{2!\lambda^{\frac{1}{N}}}\left(\frac{1}{N} \ln \lambda\right)^2 - \frac{s}{3!\lambda^{\frac{1}{N}}}\left(\frac{1}{N} \ln \lambda\right)^3 + \cdots.$$

As N gets bigger, the higher power terms of N become smaller and smaller and

$$N \ln \left(1 + \frac{s(1 - \lambda^{\frac{1}{N}})}{\lambda^{\frac{1}{N}}}\right) = N\left(-\frac{s}{N\lambda^{\frac{1}{N}}} \ln \lambda\right) = -\frac{s}{\lambda^{\frac{1}{N}}} \ln \lambda$$

since all higher power terms tend to zero as $N \to \infty$, and this term tends to $-s \ln \lambda$, so

$$N \ln \left(s(1 - \lambda^{\frac{1}{N}}) + \lambda^{\frac{1}{N}}\right) = \ln \lambda - s \ln \lambda = (1 - s) \ln \lambda;$$

that is,

$$v_1 - v_0 = -v_e(1 - s) \ln \lambda$$

in the limit as $N \to \infty$. The misleading expression in Q4, which looks like it just gets bigger as N gets bigger, actually tends to a finite velocity increment as $N \to \infty$.

7.4 Unit conversion

(a) & (b) Were solved as examples in workshop 7.4.
(c) Using 'multiply by one' method

$$\frac{9.81 \text{ m}}{\text{s}^2} = \frac{9.81 \text{ m}}{\text{s}^2} \times \frac{1 \text{ ft}}{0.305 \text{ m}} = \frac{9.81 \text{ ft}}{0.305 \text{ s}^2} = 32.2 \text{ ft/s}^2.$$

(d) Using 'convert the units' method

$$\frac{1 \text{ lbf}}{\text{in}^2} = \frac{1 \times 0.454 \text{ kgf}}{(0.0254 \text{ m})^2} = \frac{1 \times 0.454 \text{ kg} \times 9.81 \text{ N/kg}}{(0.0254 \text{ m})^2} = \frac{0.454 \times 9.81 \text{ N}}{0.0254^2 \text{ m}^2}$$
$$= 6903 \text{ N/m}^2.$$

Thus one atmosphere $= 1.03 \times 10^5 \text{N} = \dfrac{1.03 \times 10^5}{6903} \text{lbf} = 14.92 \text{ lbf}.$

(e) 1 N is the force required to accelerate 1 kg by 1 m/s^2.
Thus 1 dyne will be the force required to accelerate 1 g by 1 cm/s^2.
We can work this out in newtons using $F = ma = 10^{-3}$ kg $\times$ 10^{-2} m/s^2 = 10^{-5} N.
1 J is the work done when a 1 N force moves an object 1 m.
Thus 1 erg will be the work done when a 1 dyne force moves an object 1 cm.
We can work this out in joules using $W = Fs = 10^{-5}$ N $\times$ 10^{-2} m $= 10^{-7}$ J.

(f) Using 'multiply by one' method

$$\frac{3.00 \times 10^8 \text{ m}}{\text{s}} = \frac{3.00 \times 10^8 \text{ m}}{\text{s}} \times \frac{1 \text{ s}}{10^9 \text{ ns}} \times \frac{1 \text{ ft}}{0.305 \text{m}} = \frac{3.00 \times 10^8 \text{ ft}}{10^9 \times 0.305 \text{ ns}}$$
$$= 0.983 \text{ ft/ns}.$$

7.5 Dimensional analysis

(a) $\dfrac{\partial^2 y}{\partial x^2} = \dfrac{1}{c^2}\dfrac{\partial^2 y}{\partial t^2}$: LHS has units $\dfrac{m}{m^2} = \dfrac{1}{m}$. RHS : $\dfrac{1}{(m/s)^2}\dfrac{m}{s^2} = \dfrac{1}{m}$.

$\dfrac{\partial^2 y}{\partial t^2} = \dfrac{1}{c^2}\dfrac{\partial^2 y}{\partial x^2}$: LHS has units $\dfrac{m}{s^2}$. RHS : $\dfrac{1}{(m/s)^2}\dfrac{m}{m^2} = \dfrac{s^2}{m^3}$.

Thus the first equation is correct.

(b) Was solved in two different ways within workshop 7.5.

(c) Using the empirical method
We are trying to make a current in amps (coulombs/second).

Quantity	Unit
q	C
qu	$C \times m\,s^{-1} = C\,m\,s^{-1}$
qun	$C\,m\,s^{-1} \times m^{-3} = C\,m^{-2}s^{-1}$
$qunA$	$C\,m^{-2}s^{-1} \times m^2 = C\,s^{-1}$

(d) Using the analytic method
Let the centripetal acceleration $a = u^n/r$. Now a is measured in $m\,s^{-2}$, u in $m\,s^{-1}$ and r in m.
Thus we have $\dfrac{m}{s^2} = \dfrac{(m\,s^{-1})^n}{m} = \dfrac{m^{n-1}}{s^n}$.
In terms of metres, we have equality when $1 = n - 1$. In terms of seconds we have equality when $n = 2$. Both pieces of evidence tell us that $n = 2$.

(e)

Quantity	Unit
C	$F = C\,V^{-1}$
R	$\Omega = V\,A^{-1} = V(C\,s^{-1})^{-1} = V\,s\,C^{-1}$
CR	$F\,\Omega = C\,V^{-1} \times V\,s\,C^{-1} = s$

(f) [Electric Charge] = [Current] × [Time], thus $[Q] = [IT]$
[Voltage] = [Energy] ÷ [Charge] = [Force] × [Distance] ÷ [Charge],
thus $[V] = [FLI^{-1}T^{-1}]$, where we have used $[F]$ to represent the dimensions of force.

Now, [Force] = [Mass] × [Acceleration], so $[F] = [MLT^{-2}]$ and with this we achieve:

$$[V] = [MLT^{-2} \times LI^{-1}T^{-1}] = [ML^2T^{-3}I^{-1}].$$

Now in the SI units we define charge, voltage, and force such that the 'dimensional' equations above hold exactly without the need for extra constants. Thus we define $1\,\text{C} = 1\,\text{A} \times 1\,\text{s}$, $1\,\text{V} = 1\,\text{J} \div 1\,\text{C}$ and $1\,\text{N} = 1\,\text{kg} \times 1\,\text{m}\,\text{s}^{-2}$. Accordingly $1\,\text{C} = 1\,\text{A}\,\text{s}$, and $1\,\text{V} = 1\,\text{kg}\,\text{m}^2\text{s}^{-3}\text{A}^{-1}$.
So, the newton per coulomb is the $\text{kg}\,\text{m}\,\text{s}^{-2} \div \text{A}\,\text{s} = \text{kg}\,\text{m}\,\text{s}^{-3}\text{A}^{-1}$, which is the same as the volt per metre $\text{kg}\,\text{m}^2\text{s}^{-3}\text{A}^{-1} \div \text{m} = \text{kg}\,\text{m}\,\text{s}^{-3}\text{A}^{-1}$.

(g) Let $[F] = [\rho^\alpha A^\beta u^\gamma]$. The dimensions of the quantities are

$$[F] = [MLT^{-2}], [\rho] = [ML^{-3}], [A] = [L^2], \text{ and } [u] = [LT^{-1}].$$

So $[MLT^{-2}] = [(ML^{-3})^\alpha L^{2\beta} (LT^{-1})^\gamma] [M^\alpha L^{2\beta+\gamma-3\alpha} T^{-\gamma}] = [M^\alpha L^{2\beta+\gamma-3\alpha} T^{-\gamma}].$

It follows by equating the powers of the dimensions M, L, and T respectively, that

$$\begin{array}{ll} M: & 1 = \alpha \\ L: & 1 = 2\beta + \gamma - 3\alpha \\ T: & -2 = -\gamma. \end{array}$$

so $\alpha = 1, \gamma = 2$, and $\beta = 1$. Thus $[F] = [\rho A u^2]$. In fact the form used by aeronautical engineers is $F = \frac{1}{2} C_L \rho A u^2$ where C_L is a dimensionless number called the lift coefficient which depends on many parameters including the shape of the wing. So we were right.

(h) The dimensions of the quantities are

$[P] = [FL^{-2}] = [ML^{-1}T^{-2}]$
$[\rho] = [ML^{-3}]$
$[\mu] = [ML^{-1}T^{-1}]$
$[D] = [L]$
$[L] = [L]$
$[u] = [LT^{-1}]$

so $\dfrac{M}{LT^2} = \left(\dfrac{M}{L^3}\right)^\alpha \left(\dfrac{M}{LT}\right)^\beta L^{\gamma+\delta} \left(\dfrac{L}{T}\right)^\varepsilon$
$= M^{\alpha+\beta} L^{-3\alpha-\beta+\gamma+\delta+\varepsilon} T^{-\beta-\varepsilon}.$

Equating the powers of M, L, and T give the three equations required.

(i) We write $\alpha = 1 - \beta, \varepsilon = 2 - \beta$, and

$$\gamma = -1 + 3\alpha + \beta - \delta - \varepsilon = -1 + 3(1 - \beta) + \beta - \delta - (2 - \beta), \quad \text{so } \gamma = -\beta - \delta.$$

Then $[P] = \left[\rho^{1-\beta}\mu^{\beta} D^{-\beta-\delta} L^{\delta} u^{2-\beta}\right] = \left[\rho u^2 \left(\dfrac{\mu}{\rho u D}\right)^{\beta} \left(\dfrac{L}{D}\right)^{\delta}\right]$ as required.

(j) For P to be proportional to L, we require the power of L to be the same as P, so our equation on the right must contain L^1. This means that $\delta = 1$, and our equation simplifies to $[P] = \left[\rho u^2 \left(\dfrac{\mu}{\rho u D}\right)^{\beta} \dfrac{L}{D}\right] = \left[\dfrac{\rho u^2 L}{D} \left(\dfrac{\mu}{\rho u D}\right)^{\beta}\right]$ as requested.

7.6 Error analysis

(a) (i) $100\% \times 0.2 \div 3.03 = 6.6\%$
   (ii) $100\% \times 0.02 \div 2.34 = 0.85\%$
   (iii) $100\% \times 0.24 \div 24.3 = 0.99\%$
   (iv) $100\% \times 3 \times 10^{-22} \div 1.602 \times 10^{-19} = 0.19\%$.

(b) Standard deviation is 1.41 mV; mean is 35.14 mV.
   So we shall take 1.41 mV as our absolute uncertainty, and calculate the relative uncertainty as $100\% \times 1.41 \div 35.14 = 4.0\%$. To one significant figure the absolute uncertainty is 1 mV, and the relative uncertainty is 4%.

(c) The worst case scenario is if both heights are over estimated (or both are underestimated). This gives a total height of $7.7\text{m} + 2.2\text{m} = 9.9\text{m}$, which is 0.3 m higher than the height expected if the measurements were correct ($7.5\text{m} + 2.1\text{m} = 9.6\text{m}$). We quote our expected aerial height above the ground as 9.6 ± 0.3m.
   Thus the absolute uncertainty of the total height is the SUM of the individual uncertainties.

(d) We get the largest possible value for my daughter's mass if we take the largest value for the total mass (74.9 kg) and the smallest value for my mass (63.0 kg). We would then conclude that my daughter's mass was 11.9 kg. This is 0.4 kg higher than her mass would be if we ignored the uncertainties and calculated $74.7 - 63.2 = 11.5$kg. Accordingly we quote the expected mass as 11.5 ± 0.4kg. As with addition in (c), the absolute uncertainty is the SUM of the individual uncertainties.

(e) Redoing part (c), rather than take the absolute uncertainty as $0.2 + 0.1 = 0.3$ m, we calculate it as $\sqrt{0.2^2 + 0.1^2}$m $= 0.22$ m.
   Redoing part (d), rather than take the absolute uncertainty as $0.2 + 0.2 = 0.4$kg, we calculate it as $\sqrt{0.2^2 + 0.2^2}$kg $= 0.28$ kg.

(f) If the angle is equal to $90°$, the distance will be $\sqrt{10^2 + 20^2}$ cm $= 22.4$ cm. If the angle is less than $90°$ the distance will be less than 22.4 cm, while if the angle is obtuse the distance will be larger than 22.4 cm. Therefore we would expect the distance to average out at 22.4 cm. This implies that we would expect a 10 cm error and a 20 cm error, when combined, to average to a 22.4 cm error, even though we know that at worst the rods/errors will line up, and we will be left with a 30 cm error. This is the justification for adding errors in quadrature.

(g) (i) The uncertainty of the SUM. Here, instead of adding the nine uncertainties as $9 \times 1\,\mu\text{T} = 9\,\mu\text{T}$ to get the uncertainty in the sum, we add in quadrature as $\sqrt{9 \times (1.0\,\mu\text{T})^2} = 3.0\,\mu\text{T}$.

(ii) The uncertainty of the MEAN. The mean is equal to the sum of the measurements divided by the number of measurements (9). Given that we know the number of measurements exactly, the uncertainty in the mean will be equal to the uncertainty in the sum divided by the number of measurements, and is accordingly $3.0\,\mu\text{T} \div 9 = 0.3\,\mu\text{T}$ if the uncertainty of the sum is calculated in quadrature.

(h) (i) Expected time = $87.3 \text{ km} \div 92 \text{ km/hr} = 0.949 \text{ hr} = 56 \text{ min } 56 \text{ s}$.

(ii) Longest time = Largest distance ÷ smallest speed = $87.8 \text{ km} \div 91 \text{ km/hr}$ = $0.965 \text{ hr} = 57 \text{ min } 53 \text{ s}$.

(iii) Absolute uncertainty in the time is given by our answer to (ii) minus our answer to (i) and is accordingly 0.016 hr or 57 s. Expressed as a relative uncertainty this becomes $100\% \times 0.016 \div 0.949 = 1.7\%$ (to 2SF).
The relative uncertainty in the distance is $100\% \times 0.5 \div 87.3 = 0.6\%$.
The relative uncertainty in the speed is $100\% \times 1 \div 92 = 1.1\%$.
Thus we note that the relative uncertainty in the time is equal to the SUM of the RELATIVE uncertainties in distance and speed.

(i) Using the result we have just noted, the relative uncertainty in the resistance will be the sum of the relative uncertainties of current and voltage.
The relative uncertainty in the voltage is $100\% \times 0.03 \div 6.43 = 0.5\%$.
The relative uncertainty in the current is $100\% \times 0.1 \div 7.5 = 1.3\%$.
So we expect the relative uncertainty in the resistance to be 1.3%.
The value of the resistance is $6.43 \text{ V} \div 7.5 \text{ mA} = 857 \text{ W}$, and this is subject to a 1.3% error.

(j) When we calculate the kinetic energy we need to square the speed. This means multiplying the speed by itself, so we add the relative uncertainty in the speed to itself (i.e. multiply it by 2) to get the relative uncertainty in the kinetic energy. Given that the speed has a 8.8% error, we expect the kinetic energy to have a 17.6% error.

Let us check this. If there were no error, the kinetic energy would be 5.78 J. If the speed were at its highest permissible value (3.7 m/s) the kinetic energy would be 6.85 J. This is 18.4% higher. Thus our rule is approximately correct.

(k) First, we note that $df/dx = nx^{n-1}$. Then we substitute this into our equation for relative uncertainty to get $(1/f)(df/dx)\delta x \times 100\% = (1/x^n)nx^{n-1}\,\delta x \times 100\% = n\,(\delta x \times 100\%/x)$, which is equal to n multiplied by the relative (percentage) error in x.

(1) First we note that $\partial f/\partial x = y$ and $\partial f/\partial y = x$. We substitute these into our equation for the relative uncertainty to get $(1/f)(\partial f/\partial x)\delta x + (1/f)(\partial f/\partial y)\delta y = (1/xy)y\delta x + (1/xy)x\delta y = (\delta x/x) + (\delta y/y)$, which is equal to the relative uncertainty in x added to the relative uncertainty in y (where we have not used percentages just to keep the equations cleaner).

7.7 Locating centres of mass

(a) Each disc has a volume $\Delta V = \pi r^2 \Delta x$. However, $\dfrac{r}{x} = \dfrac{R}{X}$. So,

$$\Delta V = \pi \frac{R^2}{X^2} x^2 \Delta x.$$

The mass of this volume element is just $\rho \Delta V$. Now let us use the expression:

$$\sum_i \Delta m_i \mathbf{r}_i = \left(\sum_i \Delta m_i\right) \mathbf{R}_{cm}.$$

along the x-axis since this is a symmetry axis of our cone and its centre of mass must lie upon it. So

$$\left(\sum \rho \pi \frac{R^2}{X^2} x^2 \Delta x \times x\right) \hat{\mathbf{x}} = M \mathbf{R}_{cm},$$

which becomes an integral in the limit as $\Delta x \to \infty$

$$M \mathbf{R}_{cm} = \left(\frac{\rho \pi R^2}{X^2} \int_0^X x^3 dx\right) \hat{\mathbf{x}} = \frac{\rho \pi R^2}{X^2} \left[\frac{x^4}{4}\right]_0^X \hat{\mathbf{x}} = \frac{\rho \pi R^2 X^2}{4} \hat{\mathbf{x}}.$$

The volume of our cone is $\tfrac{1}{3}\pi R^2 X$ so $M = \tfrac{1}{3}\pi R^2 X \rho$, which means that

$$\mathbf{R}_{cm} = \left(\frac{3}{4}X\right)\hat{\mathbf{x}},$$

which is a point that is $1/4$ of the height of the cone above the base.

(b) The easiest way to approach this quite difficult problem is to calculate the centre of mass of one of the discs and this should give us a rule to locate the centres of all other discs that make up the cone. The complication of having half the cone in one material and the other half in some other material means that the centre of a disc will be located in the top half of the disc (since $0 \le y \le R, \rho = 2\rho_o$ and $-R \le y < 0, \rho = \rho_o$). Let us look at the centres of mass

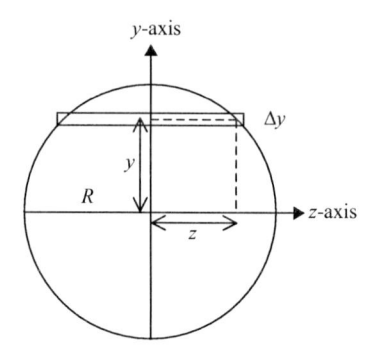

Fig. A.5

of the semicircular laminae that make up the base of the cone. The equation of this circle in Cartesian coordinates (z, y) is

$$z^2 + y^2 = R^2.$$

The centre of mass of the semicircle above the z-axis can be obtained by adding up all the contributions of mass from strips like the one shown in Figure A.5. The volume ΔV of one of these strips is

$$\Delta V = 2\sqrt{R^2 - y^2} \times \Delta y \times \Delta x.$$

Applying the general expression $\sum_i \Delta m_i \mathbf{r}_i = \left(\sum_i \Delta m_i\right) \mathbf{R}_{\text{cm}}$ we have

$$\sum 2\sqrt{R^2 - y^2} \times \Delta y \times \Delta x \times y \times 2\rho_o = \frac{\pi R^2}{2} \Delta x \times 2\rho_o \times y_{CM},$$

which becomes an integral as $\Delta y \to \infty$:

$$y_{CM} = \frac{4}{\pi R^2} \int_0^R y\sqrt{R^2 - y^2}\, dy = -\frac{4}{3\pi R^2} \int_0^R \frac{d}{dy}\left((R^2 - y^2)^{\frac{3}{2}}\right) dy,$$

which is easily integrable:

$$y_{CM} = -\frac{4}{3\pi R^2} \left[(R^2 - y^2)^{\frac{3}{2}}\right]_0^R = \frac{4R}{3\pi}.$$

The centre of mass of the semicircular lamina below the axis will also be this distance *below* the z-axis, so it is now just a question of locating the centre of the combined object the circular disc made up of a material that is twice as dense in the top half than the bottom half.

If A is the centre of mass of the top lamina and B is the centre of mass of the bottom lamina then the fact that the mass of the top lamina is greater by a factor 2 means that the centre of the circular lamina is $\frac{1}{3}$ of d away from A,

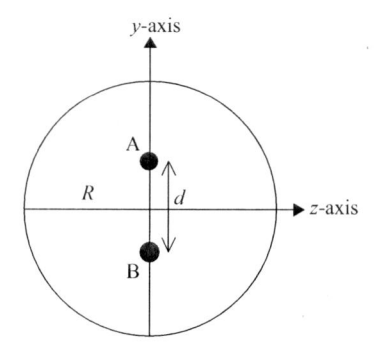

Fig. A.6

which means that the centre of mass of the circular lamina is located at y_{CM} (Figure A.6)

$$y_{CM} = \frac{4R}{9\pi}.$$

Therefore the line joining all the centres of masses of all the discs making up the cone will have an equation:

$$y = \frac{4R}{9X\pi}x.$$

7.8 Rigid body dynamics

(a) The hint tells us to have a look at the results of workshop 3.1.5 on vector triple products. You will also need to have some knowledge of *matrix multiplication*, dealt with in 3.1.2, to understand this solution. If you are less than familiar with either of these concepts please have a go at the two other workshops before attempting to go through this solution. In 3.1.5 we saw that

$$\mathbf{a} \times (\mathbf{b} \times \mathbf{c}) = (\mathbf{a} \cdot \mathbf{c})\mathbf{b} - (\mathbf{a} \cdot \mathbf{b})\mathbf{c}$$

so

$$\mathbf{r}_i \times m_i(\boldsymbol{\omega} \times \mathbf{r}_i) = m_i((\mathbf{r}_i \cdot \mathbf{r}_i)\boldsymbol{\omega} - (\mathbf{r}_i \cdot \boldsymbol{\omega})\mathbf{r}_i),$$

which can be expanded to give:

$$\mathbf{r}_i \times m_i(\boldsymbol{\omega} \times \mathbf{r}_i) = m_i \left(r_i^2 \begin{pmatrix} \omega_x \\ \omega_y \\ \omega_z \end{pmatrix} - (x_i\omega_x + y_i\omega_y + z_i\omega_z) \begin{pmatrix} x_i \\ y_i \\ z_i \end{pmatrix} \right)$$

and may be rearranged to

$$\mathbf{r}_i \times m_i(\boldsymbol{\omega} \times \mathbf{r}_i) = m_i \left(r_i^2 \begin{pmatrix} \omega_x \\ \omega_y \\ \omega_z \end{pmatrix} - \begin{pmatrix} x_i^2 \omega_x + y_i x_i \omega_y + z_i x_i \omega_z \\ x_i y_i \omega_x + y_i^2 \omega_y + z_i y_i \omega_z \\ x_i z_i \omega_x + y_i z_i \omega_y + z_i^2 \omega_z \end{pmatrix} \right).$$

Taking out the vector $\boldsymbol{\omega}$ we have that $\mathbf{L}$ becomes

$$\mathbf{L} = \begin{pmatrix} \sum_i m_i(r_i^2 - x_i^2) & \sum_i -m_i x_i y_i & \sum_i -m_i x_i z_i \\ \sum_i -m_i x_i y_i & \sum_i m_i(r_i^2 - y_i^2) & \sum_i -m_i y_i z_i \\ \sum_i -m_i x_i z_i & \sum_i -m_i y_i z_i & \sum_i m_i(r_i^2 - z_i^2) \end{pmatrix} \boldsymbol{\omega}.$$

(b) The sphere is rotating about the z-axis only then

$$\mathbf{L} = \begin{pmatrix} \omega_z \sum_i -m_i x_i z_i \\ \omega_z \sum_i -m_i y_i z_i \\ \omega_z \sum_i m_i(r_i^2 - z_i^2) \end{pmatrix},$$

as $\omega_x = \omega_y = 0$, and the sphere is symmetric, so

$$\sum_i -m_i x_i z_i = \sum_i -m_i y_i z_i = 0$$

as there will always be an element at $+r_i$ that will cancel with an element at $-r_i$ and vice versa. Therefore,

$$\mathbf{L} = \begin{pmatrix} 0 \\ 0 \\ \omega_z \sum_i m_i(r_i^2 - z_i^2) \end{pmatrix} = \begin{pmatrix} 0 \\ 0 \\ I_{zz} \omega_z \end{pmatrix},$$

where

$$I_{zz} = \sum_i m_i(r_i^2 - z_i^2) = \sum_i \rho r_i^2 \sin \theta_i \, \Delta r \, \Delta \theta \, \Delta \varphi (r_i^2 - z_i^2),$$

which is of course only an approximation until it becomes an integral in the limit when $\Delta r \to 0, \Delta \theta \to 0, \Delta \varphi \to 0$:

$$I_{zz} = \rho \iiint r^2 \sin \theta \, dr \, d\theta \, d\varphi \times (r^2 - z^2) = \rho \iiint r^2 \sin \theta \, dr \, d\theta \, d\varphi \times r^2 \sin^2 \theta.$$

(c) This integral may be performed by calculating the integrations in the three separate variables and multiplying the three results together:

$$I_{zz} = \int_0^R r^4 dr \int_0^\pi \sin^3 \theta d\theta \int_0^{2\pi} d\varphi = \left[\frac{r^5}{5} \right]_0^R \left[-\cos \theta + \frac{\cos^3 \theta}{3} \right]_0^\pi \times 2\pi = \frac{8\pi \rho R^5}{15}.$$

(d) Using determinants*

$$\mathbf{v}_i \cdot \mathbf{v}_i = \mathbf{v}_i \cdot (\boldsymbol{\omega} \times \mathbf{r}_i) = \begin{vmatrix} \dot{x}_i & \dot{y}_i & \dot{z}_i \\ \omega_x & \omega_y & \omega_z \\ x_i & y_i & z_i \end{vmatrix} = \begin{vmatrix} \omega_x & \omega_y & \omega_z \\ x_i & y_i & z_i \\ \dot{x}_i & \dot{y}_i & \dot{z}_i \end{vmatrix} = \boldsymbol{\omega} \cdot (\mathbf{r}_i \times \mathbf{v}_i).$$

So

$$K = \sum_i \frac{1}{2} m_i \mathbf{v}_i \cdot \mathbf{v}_i = \sum_i \frac{1}{2} \boldsymbol{\omega} \cdot (\mathbf{r}_i \times m_i \mathbf{v}_i) = \sum_i \frac{1}{2} \boldsymbol{\omega} \cdot (\mathbf{r}_i \times \mathbf{p}_i) = \frac{1}{2} \boldsymbol{\omega} \cdot \mathbf{L},$$

which is of course,

$$K = \frac{1}{2} \boldsymbol{\omega} \cdot \underline{I} \boldsymbol{\omega}.$$

For our sphere this collapses to

$$K = \frac{1}{2} I_{zz} \omega^2 = \frac{1}{2} \frac{8}{15} \pi \rho R^5 \omega^2.$$

(e) If $M = \frac{4}{3} \pi R^3 \rho$ then,

$$I_{zz} = \frac{8 \pi \rho R^5}{15} = \frac{2}{5} M R^2.$$

Let us assume the z-axis of the sphere is the axis it rotates about. The torque applied by F causes a rotational acceleration $(d\omega_z/dt)$ as the sphere rolls down the slope. The size of the torque is just the force (F) times the perpendicular distance from the axis (R) so it is just FR, but our vector equation tells us that

$$\mathbf{r} \times \mathbf{F} = \frac{d}{dt}(\mathbf{r} \times \mathbf{p});$$

that is, the torque is the rate of change of angular momentum $\mathbf{L}$. However, we have already seen that the magnitude L of $\mathbf{L}$ for a sphere rotating about its z-axis is

$$L = I_{zz} \omega_z,$$

so

$$\mathbf{r} \times \mathbf{F} = I_{zz} \frac{d\omega_z}{dt} \hat{\mathbf{z}},$$

---

*As before, those of you who have already met determinants will appreciate this, those of you who have not can verify that $\mathbf{v}_i \cdot (\boldsymbol{\omega} \times \mathbf{r}_i) = \boldsymbol{\omega} \cdot (\mathbf{r}_i \times \mathbf{v}_i)$ by expanding it out in full.

and
$$FR = I_{zz}\frac{d\omega_z}{dt} = \frac{2}{5}MR^2\frac{d\omega_z}{dt} \Rightarrow F = \frac{2}{5}MR\frac{d\omega_z}{dt}.$$

Acceleration down the slope is then given by Newton's second law:
$$Ma = Mg\sin\alpha - F,$$

where $Mg\sin\alpha$ is the component of the weight down the slope and F, the friction force, acts up the slope opposing the motion of the sphere. With no slip the speed of the sphere down the slope is related to the angular speed by,
$$v = \omega R \Rightarrow a = R\frac{d\omega_z}{dt} \Rightarrow \frac{d\omega_z}{dt} = \frac{a}{R}.$$

Combining all this,
$$a = g\sin\alpha - \frac{2}{5}a \Rightarrow a = \frac{5}{7}g\sin\alpha.$$

7.9 Parallel axes theorem

(a) By the Cosine Rule AC (which is the perpendicular distance of the element from the new axis) is just $AB^2 + BC^2 - 2(AB)(BC)\cos(B)$. Now $AB = d$ and $BC = r\sin\theta$, so
$$AC^2 = d^2 + r^2\sin^2\theta - 2dr\sin\theta\cos B,$$

where B is the angle between AB and BC. So everywhere we previously had r^2 (the square of the distance of the element from the axis) we need now to put in AC^2. So,
$$I'_{zz} = \iiint dM\ AC^2,$$

The mass elements are still given by:
$$\Delta M = \rho\Delta V = \rho r^2 \sin\theta \Delta r \Delta\theta \Delta\varphi,$$

and B is on the axis through the centre of mass of the sphere, which means that when we integrate over the plane containing A, B and C the $\cos(B)$ term vanishes and we are left with:
$$I'_{zz} = \rho\iiint (r^2\sin^2\theta + d^2)r^2\sin\theta\, dr d\theta d\varphi,$$

which of course collapses to
$$I'_{zz} = I_{zz} + Md^2$$

as d is a constant and $\rho\iiint r^2 \sin\theta\, dr d\theta d\varphi = M$.

7.10 Perpendicular axes theorem

(a) For small Δr the area of a ring is just the circumference around the ring multiplied by the thickness Δr. The mass ΔM of this ring is just:

$$\Delta M = (2\pi r_i \Delta r \sigma),$$

so the moment of inertia of this mass about the z-axis is $\Delta M r_i^2$, which of course means that

$$\Delta I_{zz} = (2\pi r_i \Delta r \sigma) r_i^2.$$

(b) Summing up all these in the limit as $\Delta r \to \infty$ leads to the integral:

$$I_{zz} = 2\pi \sigma \int_0^R r^3 dr = 2\pi \sigma \left[\frac{r^4}{4}\right]_0^R = \frac{2\pi \sigma R^4}{4} \Rightarrow I_{zz} = \frac{1}{2} M R^2$$

as $M = \pi R^2 \sigma$.

(c)
$$I_{zz} = \sum_i m_i r_i^2 = \sum_i m_i (x_i^2 + y_i^2) = \sum_i m_i x_i^2 + \sum_i m_i y_i^2 = I_{yy} + I_{xx}$$

The symmetry of the disc means that $I_{xx} = I_{yy}$ so $I_{xx} = I_{yy} = \frac{I_{zz}}{2} = \frac{1}{4} M R^2$.

7.11 Orbital energy and orbit classification

(a) $r = R_1 + R_2$ and $m_1 R_1 = m_2 R_2$. Solving for one say R_1 in one of these and substituting the result into the other expression leads to

$$R_1 = \frac{m_2}{(m_1 + m_2)} r \quad \text{and} \quad R_2 = \frac{m_1}{(m_1 + m_2)} r.$$

(b) $$V_1^2 = \left(\frac{dR_1}{dt}\right)^2 + \omega^2 R_1^2 = \frac{m_2^2}{(m_1 + m_2)^2} \left(\left(\frac{dr}{dt}\right)^2 + \omega^2 r^2\right) = \frac{m_2^2}{(m_1 + m_2)^2} v^2,$$

and

$$V_2^2 = \left(\frac{dR_2}{dt}\right)^2 + \omega^2 R_2^2 = \frac{m_1^2}{(m_1 + m_2)^2} \left(\left(\frac{dr}{dt}\right)^2 + \omega^2 r^2\right) = \frac{m_1^2}{(m_1 + m_2)^2} v^2,$$

so

$$\frac{1}{2} m_1 V_1^2 + \frac{1}{2} m_2 V_2^2 = \frac{1}{2} \frac{m_1 m_2}{(m_1 + m_2)} v^2.$$

(c) $$E = \frac{1}{2} m_1 V_1^2 + \frac{1}{2} m_2 V_2^2 - \frac{G m_1 m_2}{r},$$

so

$$E = \frac{1}{2} \frac{m_1 m_2}{(m_1 + m_2)} v^2 - \frac{G m_1 m_2}{r} \Rightarrow \frac{E}{\frac{m_1 m_2}{(m_1 + m_2)}} = \frac{v^2}{2} - \frac{G(m_1 + m_2)}{r}.$$

(d)
$$\mathbf{v}\times\mathbf{h} = \left(\frac{dr}{dt}\hat{x}' + r\omega\hat{y}'\right)\times h\hat{z}' = -\frac{dr}{dt}h\hat{y}' + r\omega h\hat{x}',$$

so

(e)
$$(\mathbf{v}\times\mathbf{h})\cdot(\mathbf{v}\times\mathbf{h}) = h^2 v^2.$$

(f)
$$\frac{v^2}{2} - \frac{G(m_1+m_2)}{r} = \frac{G^2(m_1+m_2)^2}{2h^2}\left((1+2e\cos\theta+e^2)-2(1+e\cos\theta)\right),$$

so

$$\frac{E}{\mu} = \frac{G^2(m_1+m_2)^2}{h^2}(e^2-1).$$

Constants

Avogadro number	N_A	6.022×10^{23} mol^{-1}
Boltzmann constant	k	1.381×10^{-23} JK^{-1}
Charge on one electron	e	1.602×10^{-19} C
Electric permittivity of free space	$\varepsilon_0 = \dfrac{1}{\mu_0 c^2}$	8.854×10^{-12} Fm^{-1}
Gas constant (ideal gas)	R	8.314 Jmol^{-1}K^{-1}
Gravitational constant	G	6.673×10^{-11} Nm^{2}kg^{-2}
Gravitational field strength at Earth's surface (typical)	g	9.807 Nkg^{-1}
Magnetic permeability of free space	μ_0	$4\pi \times 10^{-7}$ Hm^{-1}
Mass of Earth	M_E	5.977×10^{24} kg
Mass of electron	m_e	9.109×10^{-31} kg
Mass of proton	m_p	1.673×10^{-27} kg
Mass of Sun	M_S	1.989×10^{30} kg
Planck constant	h	6.626×10^{-34} Js
Radius of Earth (mean)	R_E	6.371×10^6 m
Radius of Earth's orbit (mean)	R_{SE}	1.496×10^{11} m
Radius of Moon's orbit (mean)	R_{EM}	3.844×10^8 m
Speed of light in vacuum	c	2.998×10^8 ms^{-1}

Index

absolute uncertainty, 147
acceleration
   centripetal, 56
   due to gravity, 4
   uniform, 4
   vector, 11
activation energy, 125
activation process, 125
adiabatic, 117, 119
alternating current, 105
amplitude
   complex, 106
angular frequency, 79
angular momentum, 152
   conservation, 153
angular velocity, 59
arc length, 55
argument
   of complex number, 83
atmospheric pressure, 125
Avogadro number, 129

Boltzmann
   constant, 116
   Law, 125, 127

calculus
   differentiation, 3
   integration, 8
   second derivative, 5
capacitor, 35
   in a.c. circuit, 108
carnot
   cycle, 119
   engine, 119
   theorem, 118
centres of mass, 19, 150
centrifugal force, 64
circuit Analysis
   a.c. circuits, 111
   by loop current, 104
   d.c., 102
   phasor method, 109
   using phasors, 114
closed system, 20
complex Amplitude, 106
complex conjugate, 83
complex Numbers, 81
components, 14
conic sections, 67

polar coordinates, 67
conservation
   linear momentum, 23
coordinate
   spherical polar, 137
Coriolis force, 64
Coulomb's law, 47
current
   alternating, 105
   electric, 99

dead weight fraction, 141
delta, 1
derivative
   partial, 40, 89
   second, 5
dimensional Analysis, 144
displacement, 13
distance, 5
dot product, 32
drift velocity, 99
dummy variable, 8
dynamics, 17
   of a rigid body, 152

eccentricity, 66
efficiency of a heat engine, 122
elasticity, 24
electrostatic field
   in a capacitor, 48
   of a charged wire, 47
   of a point charge, 46
ellipse, 67
energy, 32
   conservation, 116
   in a field, 32
   in a wave, 91
   in an electrostatic field, 49
   in an oscillation, 80
   internal, 116
   kinetic, 32
entropy, 123
equilibrium point, 81
error Analysis, 147

field, 29
   conservative, 42, 43
   electrostatic, 29, 35, 43, 44–9
   electromagnetic, 44
   gravitational, 29–32, 34–5, 50
   lines, 44, 45

field (*Continued*)
   magnetic, 52, 109
   non-uniform, 36
field strength, 30
flow equation, 99
force, 18
   centrifugal, 64, 65
   centripetal, 64
   Coriolis, 64
   fictitious, 64
frequency, 54
   angular, 54, 79
fridge, 118

gamma, 131
gas
   ideal, 116
   perfect, 116, 129
grad, 41
gradient function, 41
gravitational field, 50, 52, 67

heat capacity, 130
heat engine, 118
heat reservoir, 118
hyperbola, 67

ideal gas, 133
impedance, 102, 106
   matching, 92
   of a wave, 91
impulse, 22
inductor, 109
integral, 8
   line, 37
   setting up, 136
   surface, 51
   volume, 137
interference
   of a wave, 87
irreversible processes, 124
isochoric, 117
isothermal, 117

Kelvin scale, 121
Kepler problem, 66
Kepler's laws
   first law, 66, 70
   second law, 66, 74
   third law, 66, 75
kinematics, 1
   on a circular path, 54
Kirchhoff's circuit laws
   first law, 103
   second law, 103

law of falling bodies, 1, 25
line integral, 37
   evaluating, 37
logarithms, 138

magnetic field
   of a long straight wire, 37
magnetic Field, 52
magnitude, stellar, 96
matrices
   column, 11
   rotation, 56
mechanics
   linear, 1
   rotation, 54
modulus
   of complex number, 83
molecules
   diatomic, 131
   monatomic, 131
moment of inertia, 154
momentum, 17
   law of conservation, 20
multistage rocket, 141

Newton's laws
   empirical law of collisions, 24
   of motion, 20

Ohm's Law, 101
orbit
   classification, 159
   energy, 159
orbits, 66
   circular, 73
   elliptical, 73
   hyperbolic, 73
   parabolic, 73
orthogonal
   matrices, 57
   transformations, 57
oscillation
   damped, 85
   energy within, 80
   harmonic, 80

parabola, 14, 67
parallel axis theorem, 155
partial derivative, 40, 89
Path Difference, 88
payload fraction, 141
pendulum, 144
period, 54
   of orbital motion, 75
perpendicular axis theorem, 157
phase difference, 88
phase factor, 79
potential
   electric in a circuit, 100
   electrostatic, 35
   energy, 32
   gravitational, 35
power, 36
   in a.c. circuit, 106

power factor, 108
principle of relativity, 21
process
   adiabatic, 117, 119, 132
   irreversible, 124
   isochoric, 117
   isothermal, 117, 131
   reversible, 123
projectiles, 11

radians, 54
reactance, 106
reaction, 20
reduced mass, 160
relative uncertainty, 148
resistance, 106
   electrical, 101
resultant, 15
reversible processes, 123
Reynold's number, 147
right-hand screw rule, 61
rockets, 140
rotated coordinates, 56
rotating coordinate frames, 58, 62
rotating vectors, 58

scalar product, 32
semi-major axis, 66
semi-minor axis, 66
simple differential equations, 9
simple harmonic motion, 80
speed
   average, 2
   instantaneous, 3
stellar magnitude, 96
surface integral, 51

temperature, 121
thermodynamic temperature, 121
thermodynamics
   first law, 116, 125
   gamma, 131
   second law, 117
torque, 153
transformation of coordinates, 57

uncertainty
   absolute, 147
   of mean, 149
   relative, 148
unit conversion, 143

vector
   acceleration, 11
   acceleration in rotating frames, 62
   addition and subtraction, 15
   displacement, 13
   product, 58
   radius, 54
   relative, 16
   resultant, 15
   rotating, 58
   scalar product (dot product) 32
   triple product, 61
   vector product (cross product) 58
   velocity, 11
vector product, 58
vector triple product, 61
velocity
   angular, 59
   drift, 99
   increment, 141
   vector, 11
velocity distribution, 126
voltage, 100

wave
   energy carried by, 91
   equation, 89–91
   impedance, 91
   on a string, 89
   plane in, 3–d, 94
   power, 91
   spherical, 95
   travelling in, 1–d, 86
   using complex numbers, 84
   vector, 95
wave equation, 89–91
wavenumber, 87
work, 32, 117